FIREPOWER

By the same author

Passchendaele: The Tragic Victory of 1917
Kitchener: The Man Behind the Legend
Horrocks: The General Who Led from the Front
The Secret Forces of World War II
The Special Boat Squadron
Auchinleck: The Lonely Soldier
The D Day Landings
Invasion Road
Panzer
Alamein
The Zeebrugge Raid
Castles in Britain
The Soldier: His Daily Life through the Ages
Dervish
The Crimean War
The Special Air Service
The Medieval Castle

FIREPOWER

From Slings to Star Wars

Philip Warner

GRAFTON BOOKS
A Division of the Collins Publishing Group

LONDON GLASGOW
TORONTO SYDNEY AUCKLAND

Grafton Books
A Division of the Collins Publishing Group
8 Grafton Street, London W1X 3LA

Published by Grafton Books 1988

British Library Cataloguing in Publication Data

Warner, Philip
Firepower : from slings to star wars.
1. Arms and armor——History
I. Title
623.4′09 U800

ISBN 0–246–13021–0

Photoset by Deltatype, Ellesmere Port
Printed in Great Britain by
Robert Hartnoll (1985) Ltd, Bodmin

Contents

List of Illustrations

Note on Measurements

In the recent past military measurements have frequently changed from imperial to metric. In addition, certain countries have used the metric system ever since its inception. This book has retained original measurements wherever appropriate, and the conversion tables below may be of assistance to readers more familiar with one system than with the other.

Imperial to metric

1 oz = 28 g	1 inch = 25 mm/2.5 cm
1 lb = .454 kg	1 foot = .305 m
1 ton = 1016 kg	1 yard = .92 m
	1 mile = 1.6 km

Metric to imperial

1 kg = 2.2 lb	1 mm = .04 inch
1 tonne = 2205 lb	1 cm = .4 inch
	1 m (metre) = 3.3 ft
	1 km = .62 mile

Introduction

'Power grows out of the barrel of a gun.' Although many of the thoughts of the late Mao Tse-tung are now held in small esteem, this comment on modern weaponry has the indisputable ring of truth. It is not, however, the whole truth: there are many other methods by which power may be exercised. On the battlefield the power of all weapons is qualified by the tactics, leadership, training and logistical support of those who use them. An army may have massive superiority in conventional or even nuclear arms, but may be defeated because the opponent has greater morale, perhaps linked to a religious conviction which makes its soldiers regard death in battle as a passport to eternal life. Examples of the latter include Japan in the Second World War, and modern-day Iran. Every nation's individual firepower, in fact, includes a very wide range of specific factors. Thus Britain's firepower has always been related to its being an island, Russia's to the enormous distances which protect its interior, and China's to its huge population. Firepower does not even necessarily imply using guns at all, but is a convenient and widely understood term for military resources. This book examines the development of firepower, compares and assesses its various forms, and reaches some surprising conclusions.

The development of weapons has progressed from the simple club of prehistoric man to the sophisticated laser-controlled missile of today. These extremes seem to have little in common, but in fact they both derive from the same motivation – survival. Primitive man needed something to overcome the odds against him when confronted with the speed and agility, the sharp claws and teeth, and the size and protective covering of his adversaries. He was apparently totally vulnerable, but he had one enormous advantage – a higher degree of intelligence than that of his potential opponents. Had he not possessed this invaluable asset he would no doubt have become extinct many thousands of years ago. Early man knew that the main threats to his continued existence were starvation and other predators. Rivals for available food supplies might eventually become part of the food which averted starvation. Primitive hunters therefore made a triple gain: they staved off danger to themselves, and they provided themselves not merely with food but also with the assets of the dead opponent – its fur, its bones (which could be converted into knives) and its teeth, which could make formidable weapons if bound by strips of leather to a wooden spear.

The modern scientist is equally concerned with survival, although he may not realize it. Fending off starvation may be far removed from his mind, but if he has the smallest knowledge of military history he will know that many wars have quickly produced near-starvation for the losers, and sometimes, when the struggle has been protracted, destruction has been widespread and disease has taken its toll on the victors too. Therefore by designing his defensive weapon (all modern weapons are claimed to be defensive) he may well be saving himself and his country from the effects of famine. One does not need to go back as far as the Thirty Years' War of 1618–48, after which packs of wolves roamed parts of Germany, to be made aware of the relationship between famine and war; examples can be found in more recent times. The German U-Boat campaign brought British food supplies to a critically low level in the First World War, and did so again in the Second. Hitler's invasion of the Soviet Union in 1941 led to severe food shortages in Russia, but at the end of both wars it was Germany which was suffering the most. The Nazis had believed that the firepower of their Panzer divisions would easily sweep all opponents from their path. For a while they did, but the combination of Soviet T34 tanks, apparently unlimited manpower and some very efficient artillery tipped the scale against them. The Duke of Wellington is alleged to have once said there was only one thing worse than winning a battle, and that was losing it; the twentieth century has shown that, although the fate of the vanquished is usually appalling, the situation of the victor has seldom been enviable. Destroying his opponent has usually cost him dearly. One imagines that primitive man did not escape unscathed in his victories and, like the modern victor, must have wondered how many more wounds he could endure before being too weakened to take on further opponents. The situation after another global war may easily be visualized: contaminated food supplies, polluted waters, destruction beyond the power of survivors to repair, and widespread disease. The modern military scientist, if he cares to contemplate the fact, is in the same situation as his primitive forebear. He is not, of course, working in order to eat his opponents but to ensure that they do not destroy his own food supplies. The weapon he is working to create is meant as a deterrent; primitive man used similar steps to scare away inedible wild animals which threatened his home and food supplies.

On 22 November 1986 *The Times* reported that sixty thousand refugees from northern Afghanistan had entered Pakistan, bringing the number of refugees in that country up to three million. The latest batch, half of them children, said they had been driven out because of systematic crop destruction by the Soviet occupation forces. Survival through the winter had become impossible. All forms of siege, whether of castles, towns, countries or even resistance groups, have always used starvation as a valuable, even primary, weapon. Regardless of the ideological or other objectives with which wars may begin, they quickly settle down to policies of attrition and survival. This has been true from the days when primitive sieges reduced fortresses by starvation to the present century: in both the First and Second World Wars Germany nearly brought Britain to submission by U-Boat warfare directed against supplies crossing the Atlantic, and Russia hampered the German armies by what became known as the scorched earth policy, denying the enemy the chance of 'living off the land'. Perhaps the greatest deterrent to a Third World War is the fact that the potential belligerents are only too well aware that modern weaponry, even short of nuclear, would rapidly reduce the situation to a struggle for the remaining inadequate food supplies.

1

Pre-Gunpowder

Many weapons used by primitive man are still, unfortunately, in use today, and are often lethal. For a person who has never been the target, a skilfully thrown stone may seem a relatively harmless weapon, but this would not be the view of police or troops facing hostile crowds: even a riot shield does not give complete protection. Primitive man would have viewed with contempt the lack of force and accuracy shown by the modern missile thrower – he himself would have hurled his jagged stone with considerably more power than the average rioter. He may perhaps be compared with today's fast bowlers on the cricket field, who project a 5½-oz ball at a speed of over 90 miles an hour. The ball itself, consisting of cork and twine covered with smooth red leather, is not intrinsically lethal, but when aimed at a target some 20 yards away it can kill, even though it may have bounced once on the ground before reaching its victim. Sharp flints which had doubtless been used as throwing stones may often be found in fields where arrowheads and a nearby burial mound indicate some ancient conflict. Hundreds of years later throwing axes were found to be even more deadly, as well as being easier to control than flung stones. The boomerang, when used by an Australian aboriginal, is, of course, a killing weapon.

One of the most impressive accounts of the latent power of the small stone comes in the biblical story of David and Goliath. David carefully chose smooth stones for his sling, a weapon which dates back to very early times. Their smoothness would ensure accuracy, for when aiming at the armoured giant confronting him it was vital to hit one of the few unprotected places. One blow was sufficient: Goliath fell to the ground, unconscious or already dead, and David hacked off his head to make sure of his victory.

Slings were used widely in the ancient world. Just as later certain peoples specialized in the use of a particular weapon (Genoese crossbowmen, Welsh archers, Saxon axemen), so did the sling have its élite troops, who often hired themselves out as mercenaries. The slingers of the Balearic Islands in the Mediterranean were regarded as more formidable than archers. Long after more sophisticated weapons had been invented and come into use, the sling was still viewed as an awesome weapon. Ammunition was so plentiful that a target such as castle battlements could be blanketed with stones, impossible to dodge as they came curling down like a shell from a howitzer. Eventually the sling drifted into disuse, not because it was obsolete, but because it became impossible to obtain slingers with the necessary skills: it was easier to train an archer than a slinger, and many fewer archers than slingers were needed on a battlefield.

As every soldier knows, the efficient use of a weapon depends on the training of the user. Early man had certain advantages: he was stronger, more agile and faster than his modern counterpart, and his eyesight was undoubtedly superior to ours. The Bushmen of the Kalahari Desert today can see objects at a distance for which Europeans would need powerful binoculars. Furthermore, the Bushmen have a strong visual memory. They are unable to count above three, but if they see the number-plate of a passing car they can draw the figures in the dust with a stick, without any idea of what they signify. Primitive man was highly trained in the weapons he possessed, and the constant need to use them in order to survive ensured that his skills never grew rusty. Later the Roman soldier trained for eight hours every day, and the English archers of the Middle Ages were required to practise to the point of exhaustion. Once cavalry came on to the battlefield, training took an even more complex form: by the nineteenth century two years were needed to train a lancer in the Indian Army. The British marksmanship of the 1914 regular army was so rapid and accurate that it was mistaken by the Germans for machine gun fire. Such standards are not needed today with modern automatic rifles, but some infantry weapons are now so sophisticated that intensive training is necessary if they are not to be misused and thus made ineffective.

One reason why so much is known about primitive weapons is that in recent times specimens have been found in peat bogs or gravel pits, where the surrounding material has kept them close to their original condition. Early spears were formidable weapons, sometimes tipped with flint, sometimes with points hardened by heating, though not burning. Later, when they were tipped with bronze or iron, they became even more fearsome. They were, of course, much heavier than the modern sports javelin. One reason why men were often wounded or killed by spears or arrows is that an approaching spear is notoriously difficult to judge. It appears as if it is going to fall some yards short of you, but to your surprise and, perhaps, dismay, it will come much closer. Many a man has been wounded by a spear which he did not think it was necessary to dodge. Today accidents occasionally occur on sports fields when spectators of javelin throwing have approached to what they thought was a safe distance, only to find out their mistake. Formidable though the thrown spear is, it was not formidable enough for Chaka, the Zulu chief, in the early nineteenth century. He insisted that his warriors should close with their opponents and strike at their targets with a hand-held spear. Chaka's warriors were, of course, using weapons and equipment which differed little from those used by other fighting men some three or four thousand years earlier. Their tactical skills were, however, much more advanced.

A valuable addition to early man's firepower was the bow. For centuries it remained a weapon with a short range and was thought to have remained so until the medieval period, but a few years ago a prehistoric bow discovered in a Swedish bog appeared to have many of the characteristics of the longbow of the Middle Ages. The bow, as will be seen later in the course of this book, was destined to have a sophisticated future, taking many forms, using many different types of arrow, and bringing into battle elaborate tactical measures. Early arrows were effective enough when plentifully used, and no doubt early man, like certain jungle dwellers today, treated the points with a variety of poisons. The form of poison used was largely determined by what was easily available. The best poison for the hunter was a substance which paralysed the nerves of the victim but did not contaminate the flesh. This might be similar to curare, used in South and Central America, which is derived from plants of the Strychnos group. Alternatively, if needed against dangerous but inedible foes, a form of poison would be used which would kill the recipient almost immediately. Such poisons might well be derived from snake venom. Poison was used not merely on arrows but also on darts which, propelled from a blowpipe, would be silent, unseen and totally lethal. For certain purposes darts are the ideal weapon; it is not very relaxing to hear a bullet hiss as it passes close to your head, but the gentle sigh of a dart from a blowpipe brings a totally different category of fear.

Examples of primitive weapons may be found in many museums today: arrows, spears, shields and flints are widespread. However, seeing them on display does not give close enough acquaintance to judge their effectiveness. To test the cutting edge of a flint is an interesting experience: it is usually surprisingly sharp, even though some fifty thousand years may have passed since it was first chipped. And, of course, the potential of any weapon can only be judged by seeing it handled by a strong, skilled and determined user. I have seen both modern Indonesian tribesmen carrying spears tipped with poison, and aboriginals in northern Thailand armed with blowpipes.

Early man's weapons never left his side, day or night, and when he died, if the circumstances of his death made it possible for him to have a grave, they were buried with him. For whatever he encountered in the unknown, he must have his weapons with him.

Although primitive men had hunted in groups or tribes when seeking the larger animals, it is thought that inter-human warfare did not begin until much later. They had little reason to oppose each other: they were not competing for space as there was clearly sufficient territory for all, and the obvious enemies, such as predatory animals, could best be dealt with by combining and fighting as a group. Later, when the population had grown larger, when agriculture made certain areas more desirable than others and when certain hunting grounds had been denuded of game, it was a different story. So it may be assumed that throughout the Palaeolithic period (the Old Stone Age), which lasted from 500,000 BC to 9000 BC, man did not turn his hand against man, and his weapons were used against animals only. In the next period, the Mesolithic (the Middle Stone Age), which lasted from 9000 to 4000 BC, there is uncertainty about man's attitude to his neighbour, but from the Neolithic Period (the New Stone Age, 4000–1500 BC) there is evidence that men were developing warlike tendencies.

The period brought a wave of invaders into Europe – there had been earlier ones from northern Africa in the Mesolithic – but now there was less unused space and more people to occupy it. From this Neolithic period date causeway camps, usually on hilltops, with two or three rows of ditches round them. It was long thought that their design made them too vulnerable to be used against anything but animals, but excavations at Crickley Hill, near Cheltenham in Gloucestershire, revealed a Neolithic camp which had been attacked, captured and burnt. The area was strewn with arrowheads; radiocarbon dating placed this battle at approximately 3000 BC. The battle of Crickley Hill was therefore fought a thousand years before the building of Stonehenge, and the fact that the only weapons which we can prove were used in the battle were arrows does not rule out the possibility that more sophisticated ones may also have been employed. The people who lived around Stonehenge were advanced in many ways: they wore woven clothing, they used well-designed implements for eating and drinking, and they had some success with surgery, as can be judged from mended bones.

The Bronze Age (approximately 1600–700 BC) brought impressive new power to weaponry. Bronze is an alloy made predominantly of copper, with some tin; it can be fashioned to make sharp knives and spearheads, light but lethal arrow tips, and strong shields. The people who used these weapons and built the round barrows which may still be seen dotted over Salisbury Plain and at places west eventually moved into Wales, perhaps because they were driven there or perhaps because they found the Welsh coastline more fertile.

The eras of weapon development were now coming more closely together. Around 700 BC, or possibly even some two hundred years earlier, the Iron Age arrived in Britain. This was the period of the great hill forts, and the remains of some two thousand of them can still be found in Britain. They were built with primitive iron tools and unlimited labour. As an average person's lifespan at this time was less than thirty years, it must have been one of incessant toil. Deep trenches were hacked out of tenacious clay: in some forts this meant scraping away 30,000–40,000 tons – a frustrating, exhausting task. Treetrunks were cut, shaped and built into the ramparts by the thousand. Today only worn and weathered remains can be seen, but those are impressive enough. What these forts must have looked like when new, and complete with defensive superstructure, can only be guessed at, but they must have been a formidable sight. And if these were the defence, what must the attack have been like, and how armed? Certainly one weapon used with great effect on these fortifications in the Iron Age, and which has never gone out of use, was fire itself.

Fire was launched at wooden, and even stone, defences in a variety of weapons. Several types of combustible material, varying from pine resin to compounds of pitch, were well known. Thus the materials which helped to keep ships afloat were also ideal for ensuring their destruction. Fire made short work of brushwood and thorn, which otherwise made ideal obstacles to advancing soldiers. Such obstacles needed to be constantly drenched with water if they were not to be quickly burnt away, but in a hill fort there was rarely enough water for this purpose in the summer months. Sometimes the brushwood was covered with wet hides but these, while protecting it, made it easier to cross. When William I of England was engaged in one of his innumerable sieges in Normandy the inhabitants put out wet hides on the ramparts, calling out as they did so, 'Hides for the tanner.' The reference was to William's father-in-law, a tanner by trade, and William did not miss the insult: tanners were considered to have the most noisome and unpleasant of all occupations and to smell accordingly. When the town eventually fell, William rewarded the inhabitants for what he felt was a misplaced sense of humour by chopping off their feet and throwing them off the walls.

Fear of being burnt alive or defeated drove men to ingenious measures. In the twelfth-century siege of Exeter, the inhabitants ran out of water and used wine to put out fires which were starting on the battlements. In the following century, at the siege of Château Gaillard, near Rouen, fire was used extensively by both attackers and defenders. A contemporary writer describes the exploits of an early frogman who filled some pitchers with live coals, sealed them in and tarred over the top to make them waterproof. He then swam down the Seine in the darkness, towing the pitchers behind him, and came out of the river below the castle walls. At this point there were buildings and wooden defences along the narrow strip of shore, but they were now unoccupied since the former occupants had withdrawn within the walls and were engaged with the enemy who were attacking on the far side. The frogman used his unusual cargo to set light to the wooden outbuildings, thereby creating much confusion, smoke, alarm and diversion of force. Helped by an east wind, the frogman's fire gave the impression that the whole of one side of the fortress was ablaze.

The most ingenious uses of fire came, not surprisingly, from the Middle East, where petroleum and its by-products were easily found. One of the most effective was what came to be known as Greek fire, although it probably did not originate in Greece. There may have been several different forms of Greek fire, but the one fact common to all is that no one today has any idea of the formula by which it was made. Phosphorus may have been one of the ingredients, for this chemical was known to burn fiercely on water. Sulphur would have assisted Greek fire to stick to stone surfaces, and pitch would help to keep it burning, but the reason for its explosive qualities can only be conjectured. It could be delivered by pigeons which were unaware they were on kamikaze missions or, even more mysteriously, it could be blown from tubes, not unlike a modern flame thrower.

Ordinary fire was also employed indirectly by defenders of towns and castles to heat up sand. A bucket full of red-hot sand emptied over men scaling a ladder would penetrate between the joints in armour and produce appalling devastation. (Even more feared was quicklime which, scattered down-wind, would blind the unsuspecting enemy.)

In castle sieges fire could be launched over the battlements by arrow or catapult, or else wagonloads of dry brushwood could be pushed forward to the walls and then ignited. It was a favourite weapon for attacking gatehouses. The main entrance to a castle was inevitably its weakest point: it could not be made of materials which were too heavy to open, which meant that it was probably built of wood protected by a metal covering. To make up for its intrinsic vulnerability, the gatehouse was flanked by every form of defensive device which the castle-holder could provide. Above it was a projecting shelf, through the floor of which lethal objects could be dropped on would-be attackers. These rampart walls, built into a system called machicolations, were called allures. In addition there were portcullises, drawbridges and various provisions for protecting the front of the gatehouse with flanking fire from arrows.

Nevertheless, history demonstrates that the gatehouse often provided the easiest point at which to force entry. Attackers would approach carrying mounds of brushwood which had been treated to make it highly combustible. As they did so they were able to use their burdens as a shield against the arrows and other weapons which the defenders would now be launching towards them. As soon as the attackers reached a point close to the door they set alight their mobile bonfire. Although the defenders on the allure would now be pouring down water in order to quench, or at least limit, the flames, this was seldom effective: soon they themselves would be driven back from the battlements by the smoke and flames, while the great heat below weakened the door and damaged the surrounding structure. However, once the attackers reached the second line of defence they would be unable to continue using fire as a weapon: in the now confined space it would be equally dangerous to both sides. At all stages the smoke added to the confusion and made effective control of either attack or defence virtually impossible.

Fire was used not only from above, and at the front and sides, but also from below. It was particularly effective against castles with square corners. The technique was to drive a tunnel underground to a certain point beneath the walls. Having reached the required position, the tunnel would be widened to make room for a considerable quantity of incendiary material, which was usually brushwood soaked in resin and fat. Once in place, this primitive mine would be set alight: in normal circumstances it burned so fiercely (presumably there was a draught in the tunnel) that its great heat had the effect of an explosive. The corner structure would then split and fall outwards. The best-known example of this happening was at Rochester in 1216. King John displayed great determination when he set out to recapture Rochester, which had been seized by the barons who opposed him. A tunnel, of which traces remain, was started 40 yards away from the castle; when it had reached its destination it was packed with the carcasses of forty fat pigs, specially collected for this purpose. The result of lighting this underground bomb was that the whole of the south-west corner collapsed. Visitors to Rochester Castle today are able to see for themselves the results of this example of medieval firepower, for three of the corners have their original square towers while the fourth was rebuilt in a circular shape, as that form had proved to be less vulnerable to underground operations.

Attack by mining underneath walls is usually thought of as being a technique of warfare dating from medieval times, but in fact it had been in use for over a thousand years when it was employed at Rochester. In pre-Christian times, when cities were built with formidable walls, mining was a popular method of attack, if only because it was less costly in casualties than assault by scaling or battering ram. Whether under siege or not, the custodians of ancient towns and fortresses were very much alert to the possibility of surprise attacks of this nature. They placed bowls of water on the ground at intervals outside the fortress walls, and kept them under close observation: if the water was seen to be rippling when there was no atmospheric reason it was obvious that hostile action was proceeding below ground, and needed to be investigated. An additional safeguard was to place flat brass plates against the walls at regular intervals and use them as listening devices. This technique of deducing a hostile presence by sound detection would be seen again in ASDIC, an early form of sonar developed in the First World War for submarine detection; it took its name from the British Admiralty Anti-Submarine Detection Investigation Committee.

Rochester was perhaps the most literal example of firepower delivered by tunnel, but there were many subsequent examples of tunnels being used in other ways which were no less effective. The tunnel became an essential component of firepower because it was part of the delivery system – firepower should always be used at the point where it will be most damaging. Tunnels can therefore be used to deliver soldiers, explosives or even more sinister agents of destruction, and underground attack has never gone out of favour. It was used extensively throughout the sixteenth, seventeenth, eighteenth and nineteenth centuries against forts, towers and castles, and has been dramatically developed in the present century. In the First World War tunnellers were active on various parts of the Western Front: they blew up Hill 60 on April 1915 and Hooge in the following July, and produced an explosion of horrendous proportions at Messines on 7 June 1917. Messines lies on a ridge 7 miles from Ypres, and was strongly held by the Germans. Although both sides were known to be mining and countermining, the Allies managed to drive twenty deep tunnels into the hill beneath the German position. Nineteen of the mines were exploded with devastating results; one, now somewhat ironically called the Pool of Peace, contained 91,000 lb of ammonal which blew a hole over 200 feet wide and 40 feet deep. And that was only one. The combined explosion was heard in the middle of London. The Allied mining operation was undetected because their mineshafts were driven in at a much lower level than the Germans had expected. The latter were totally surprised, although they had listening devices which were highly sophisticated by previous standards.

In the Second World War underground tunnels were often used for offensive/defensive warfare rather than as shelters from bombardment. On strategic islands, such as Iwo Jima, the Japanese constructed an elaborate system of underground passages which were designed to enable them to attack an oncoming enemy on the flanks or from the rear.

In more recent times two events have shown that offensive by tunnel is by no means relegated to the past. In the Vietnam War the North Vietnamese showed themselves adept at burrowing under American positions: considerable surprise was expressed when *The Times* published a picture of a soldier using a doctor's stethoscope to listen for any hostile movement below his feet. A more unusual use of a tunnel to enhance one's own progress, and to detract from that of the opposition, was the Anglo–American tunnel in Berlin in 1954. The aim was to drive a 1500-foot-long tunnel under the boundary between East and West Berlin, from which most of the messages passing to the Red Army and elsewhere could be tapped. Unfortunately for NATO, the spy George Blake betrayed the existence of the tunnel to the Russians. The Berlin tunnel was an example of adapting an ancient means of penetrating the enemy citadel; on this occasion it was used not to destroy, but to discover the potential enemy's intentions.

The tunnel even became a weapon in its own right. Traps consisting of holes covered with an easily broken camouflaged surface had, of course, existed for thousands of years, and been very successful in immobilizing unwary animals or humans. In the Middle Ages the tunnel/trap, originally thought of as a defensive weapon, was soon to be used for positive, destructive action.

One method of attacking castles used from very early times was to construct mobile siege towers, fill them with armed men and then have them pushed forward to the walls of the fortress under attack. As they came close a drawbridge would be lowered from the tower on to the battlements, so that the attackers could rush in at that point. The defenders found it very difficult to fight off the medieval tank-like object, as it possessed very considerable firepower of its own. It was usually protected with wet hides in case a lucky incendiary arrow should fall on a suitable target and set it alight. But the formidable monster was soon discovered to be easily stoppable as long as the ground over which it advanced was neither rocky nor marshy. The defensive technique was to dig a tunnel or series of tunnels at some distance from the fortress walls, so that they completely encircled the buildings. The roofing of these tunnels was strong enough to bear the weight of a limited number of foot soldiers, or even horsemen, but when the great weight of a siege tower went over it, it would collapse. The attacker was then at his most vulnerable. The siege tower usually fell sideways, injuring some of its occupants in the process; from this moment it was immobilized and helpless, crammed with men, and an easy target for the incendiary arrows which would now be directed on to it. The concealed tunnel had been the prime agent of destruction.

The weight which the tunnel roof would bear was carefully calculated, and by no means a matter of chance. When siege towers were not expected, probably owing to lack of suitable materials from which to construct them, the tunnel traps were adapted for attackers on foot. For them the roof would be weaker and the base of the tunnel would be lined with sharp spikes; the height of the tunnel would be such that, once the victims had fallen in, there was little possibility of their climbing out again.

In one of the most recent wars of the present century – Vietnam – the tunnel-trap was used by the Vietnamese to break up attacks and therefore reduce the firepower of the oncoming force. Such techniques are sometimes known as passive defence, but there is nothing passive about tunnel/traps when employed on battlefields.

Whatever the method of attack or defence, fire itself was always likely to be used. It is, of course, a terrifying weapon. Few boys grow up without reading of the battles between American Indians and settlers in the nineteenth century. The settlers, ever wary of attack, protected themselves by wooden stockades – wood being their only building material – within which lay their living quarters. The Indian braves, with no regard for their own survival, would approach carrying burning torches and hurl them into the besieged fort. Although many Indians were shot, the survivors were able to pick easy targets among the settlers who were desperately trying to put out the flames and were clearly outlined by them.

But the primitive weapon of fire achieved its greatest efficacy in the Second World War. The incendiary bomb had limited uses on the open battlefield, but against cities, ports and shipping it was devastating. Incendiaries delivered by aircraft were scattered over wide areas, and the fires they started made the task of the following bombers easier. Experience soon showed that it was an effective policy for a pathfinder pilot to drop incendiaries on a target some distance to the right or left of the main objective. The defence closed in to the spot, which they assumed was the target for the raid, thus leaving the main objective of the attackers relatively undefended. This procedure was always carefully planned before the raiders set off, and everyone therefore knew where to go.

Bombing with incendiaries was, of course, practised by both sides during the Second World War. The Germans used them to great effect over Britain in the latter months of 1940. After the fall of France, the Luftwaffe began a series of day and night raids on British cities, notably London. However, the day battles of September, culminating in the unusually large one on 15 September, convinced them that daylight raids were too expensive. From that point they abandoned attempts at precision bombing and, instead, set about destroying large areas with a mixture of high explosive and incendiaries. The city of London suffered heavily in these raids, which became known as the Blitz, from the German word for lightning.

On 22 June 1941 the war took a new turn: Germany suddenly invaded the Soviet Union, a country with which it had a non-aggression pact. After a brisk start the German attack slowed down; the terrain and the weather were useful adjuncts to Soviet resistance. But the factor which impeded the Germans most was not the vast distances which needed to be covered, but the fact that the Soviets had found a new use for the fire weapon – it became known as the scorched earth policy. Every place and object which might be of value to the invading Germans was systematically destroyed by the retreating Russians. Buildings and crops were burnt, and everything which could not be carried away was destroyed. Although it came as a surprise to the Germans, this was not the first time the Russians had used fire in this way. After Napoleon launched his invasion of Russia on 24 June 1812 he, too, made slower progress than he had hoped for. Nevertheless, he entered Moscow on 14 September; five weeks later he had to abandon it and begin his long, disastrous retreat. He had found the city in flames, which burnt for a week. Whether the fire had started accidentally or not, no one ever knew, but for Napoleon its result was catastrophic. For a while he wondered whether to try to press on to St Petersburg (Leningrad), but realized that that might bring him into even greater trouble.

The policy of leaving scorched earth has now become an accepted military technique. But before the end of the Second World War another form of destruction by fire had reached a stage of horrific efficiency which seems barely credible. Incendiary bombing in Europe was appalling enough, but when the United States Air Force bombed Tokyo and Yokohama in the spring of 1945 they are believed to have killed more than the combined total of deaths in Hiroshima and Nagasaki. Yet, curiously enough, the fate of the unfortunate occupants of Japanese cities which were subjected to conventional bombing has never received a fraction of the publicity given to those cities destroyed by nuclear weapons. It is, of course, true that the after-effects of the Hiroshima and Nagasaki bombs, such as radiation sickness, were a new and terrifying aspect of warfare, but disease following destruction had been a familiar phenomenon after many previous wars. Among the more sinister of weapons in the stockpile of certain nations are those classed as 'bacteriological and chemical'. They include, no doubt, germs of the most notorious killer diseases known to man, such as anthrax and bubonic plague, but it will be a rash nation which first decides to use them. The spread of disease seems to be well beyond man's control, and the nation which first uses a germ as a weapon could eventually turn out to be the greatest sufferer. One of the most destructive of post-war diseases rampaged in 1918–19 – the ill-famed, and probably wrongly attributed, Spanish flu. At the time it was thought that the rapid spread of this devastating epidemic, which killed nearly twenty million people (a figure exceeding the total war casualties), was the result of the debilitation which had been the experience of many during the war years, but, confusingly, it also spread into countries which had remained untouched by the war.

So unpredictable are the effects of certain weapons of mass destruction that only a degenerate or totally irresponsible country would resort to them, however much these weapons might appear likely to produce a rapid and complete victory. History shows that governments are extremely cautious about trying out new weapons. The use of nuclear bombs against Japan, for instance, was only sanctioned because it was known that to conquer Japan with conventional weapons would probably result in a million Allied casualties, among whom would be all the prisoners of war held by the Japanese.

When an available weapon is not used, there are two possible reasons. One is that the effects are too unpredictable – it might rebound on the user or make conquered territory untenable. The second is that it has been banned by the belligerents at a previous international conference.

Attempts to make war less destructive have existed for at least a thousand years. Usually, but not invariably, those pressing for a ban have done so for military, rather than moral, reasons, although the two have often been confused. The success of the crossbow in Europe in the eleventh century caused the Lateran Council at Rome, one of the five councils of the Western Church, to ban this weapon in warfare fought by Christians. The reason given was that it was too devastating to be used by the people of Christian nations. The crossbow was certainly a formidable weapon, though not an entirely new concept: earlier, larger versions had been used in pre-Christian warfare. It was, of course, terrifyingly powerful and, to some extent, unpredictable, but it seems that it was banned not so much for its destructive capabilities as for its indiscriminate nature. When used against opponents some five hundred paces away, it was not possible to say who would be the recipient of the bolt; it might even be a king and it could easily be an earl or an important knight. Medieval warfare had its own special rules, and they greatly favoured those who organized it. Kings, earls and knights did not go on to the battlefield to be slaughtered: they had horses on which they would be raised above the common throng, and on which they could make a hasty departure if all was not going well with their supporters. They also wore armour which shielded them from harm while they were inflicting wounds on those less well protected. If they encountered another person of equal rank, the facts of the matter would be known to both, the due procedure of chivalry would be observed, and an unhorsed opponent would probably be helped to his feet and preserved for his ransom money. But neither the crossbow nor its successor, the longbow, took account of this chivalric tradition. Its evil bolt flashed through the air finding a target at random, and meeting no serious impediment in the chain mail of the period. The unfortunate recipient was hurled to the ground, wounded and perhaps dying, and foot soldiers who were more concerned with winning than with social distinctions would slip a dagger through a convenient joint in the armour.

Banning the crossbow did not, of course, prevent its use, and soon all armies had their squadrons of élite crossbowmen and longbowmen. In consequence another form of deterrent was introduced – stronger and thicker armour, usually plate, which would stop a normal bolt or arrow. But even plate did not make the wearer invulnerable for long. Soon the crossbow bolt, now called a quarrel (it was four-sided, as in 'quadrangle'), had four spikes specially designed to crack open armour and leave the wearer exposed to the next shot.

Proscribing a weapon because it is too horrible to use seems only to be effective if its use may bring into action an even more terrible weapon in retaliation. At the time of writing, British farmers are pressing for laws which will require crossbow owners to hold the same licences as those issued for firearms. The reason for this current dislike of the crossbow is that it is used by poachers to kill deer, cattle and sheep. However, those crossbow owners who intend to use their weapons in order to steal somebody else's property are hardly likely to request a licence from the police to enable them to do so, and although we may perhaps soon see a ban on the sale of crossbows, this is likely to be no more effective than the existing laws on the possession of firearms; the only people who apply for licences to use powerful weapons are those who have no intention of using them for criminal purposes. However, as will be seen in the course of this book, it has been, and is, possible to restrain the use of certain weapons if it seems to be to the advantage of all sides.

The crossbow is only one of several varieties of bow, each of which has its special idiosyncrasies. All, however, share certain characteristics: they are cheap to manufacture, deadly in effect, virtually silent, and require practice and skill for effective use. The crossbow is the heavyweight of the species, and up until the present century was generally accepted as having double the range of the longbow. The records for both are currently held by Americans, for the crossbow with 1243 m and for the longbow with 1122.25 m; the smallness of the disparity between the two shows that the superiority of the crossbow over the longbow has now been almost entirely eroded. However, to compare the distances achieved by weapons using modern materials and made for competition with those achieved by weapons of war, which have to conform to other specifications, is not entirely satisfactory.

The principle difference between the two forms of bow was that one was entirely hand-powered, while the crossbow used a mechanical aid – either a lever or a form of winding mechanism to draw back the bowstring, which was then released by a trigger. As with many other inventions, the first record of crossbows came from China, in 210 BC. Drawings, usually on tomb reliefs, and archaeological discoveries of parts of crossbows, suggest that to have reached such a degree of development they had already been in use for several hundred years by the time those records were made.

The principle of the crossbow was adapted into the manufacture of the much larger weapons which became the forerunners of siege guns. This artillery type of crossbow was more popular with the Romans than the hand-held variety, but it did not lead them to neglect the possibilities of the smaller weapon entirely. Records show that the Romans occasionally used crossbows in battle, but it seems that they were too slow to fit in with Roman close-quarter fighting. Then, and later, the crossbow seems to have been more popular for hunting than on the battlefield.

The range and power of the crossbow made it ideal as an early form of sniper's weapon. The archer did not need to stand up to draw it: he could wind up the string and then crouch concealed in a suitably camouflaged position. When his victim came into view, careful aiming and gentle pressure on the trigger were all that was required. It is said that when William II of England, known as William Rufus, was killed in a hunting accident in the New Forest in 1100 there was some doubt about the nature of his death because he was transfixed by a crossbow bolt, which may or may not have arrived accidentally. (Hunting 'accidents' were by no means uncommon: the chase offered numerous opportunities for despatching rivals.) The crossbow was an assassin's weapon, and an ideal instrument for revenge killing. Richard I of England, the absentee French-speaking king who became a British national hero, used the crossbow in defiance of the views of the Roman Catholic Church but ironically was himself killed by a crossbow bolt; it was aimed by a man who accused Richard of having killed three members of his family. Crossbows were not liked, even by those on their own side. You needed to be a man to stand up and pull 70 lb on the string of a longbow, but any ill-favoured varlet could wind up a crossbow and prop it into an aiming position.

A weak point of the crossbow was that in wet weather the string could not be removed and kept dry, whereas a longbow string could be slipped off and curled inside the archer's waterproof hat. At the Battle of Crécy in 1346 the squadron of Genoese crossbowmen, mercenaries who were fighting with the French Army, were made ineffective by a sharp shower of rain which slackened the tension of their bowstrings. The French knights were so infuriated by the failure of the cross-bowmen from whom so much had been expected that they rode through them and slashed them with their swords. Suspicion and resentment of specialists using unorthodox weapons is common in most armies, even today, but feelings are not normally expressed as strongly as they were at Crécy.

Nevertheless the crossbow never went out of fashion, although its later uses tended to be against animals rather than humans. Its power in relation to its size, and the fact that it can be wound up away from where it will be used, makes it an ideal weapon for use against big game such as tigers (sometimes as a spring trap), or even for spearing fish. In the Middle Ages a wooden crossbow bolt, about 1 foot long and ½ inch in diameter, was very popular for spearing river fish. If it missed – which was rare – the bolt would float to the surface and could be retrieved.

Yet another use was found for the crossbow by shaping the bolt to give it a curved cutting surface, usually in a crescent shape. This device proved very effective in slicing through the rigging of enemy ships in sea battles; thus crippled, the unfortunate victim would be unable to manoeuvre out of danger.

In its heyday the crossbow was usually known as the arbalest, a word compounded of Latin *arcus* = bow, and *ballista* = military engine. Nowadays it is not only referred to as a crossbow, but is also often miscalled an arquebus. The arquebus or harquebus was an early form of handgun.

The handbow had a longer and more distinguished career than the crossbow: it appears in Mesolithic cave paintings in Spain, which dates it to at least five thousand years BC. It is, of course, such an obviously useful, yet simple, weapon that it has been used in many different countries in many different ways. It is often heard of in pre-Christian warfare. The Assyrians, who were a major power in the area around modern Iraq and Syria and the eighth century BC, used the handbow with great ingenuity and success. They mounted their archers on heavy horses two at a time. One of the horsemen carried the bow, which was about 4 feet long, while the other held a supply of 3-foot-long arrows. As the horses had no saddles and no stirrups, horsemanship needed to be of a high order. The arrows were tipped with iron, of which material the Assyrians had a monopoly for many years. To the modern mind it would seem almost impossible to aim an arrow straight from a moving horse – it is difficult enough to do so with a rifle bullet – but the Assyrians apparently did it.

The bow is heard of again with the Parthians who, in the second century BC, built up a formidable empire in what is now Iran, and even defeated Roman armies sent against them. Their firepower lay in heavy and light cavalry. The former carried long lances and heavy bows; the latter also had a selection of arms, including swords and daggers, but their main weapon was the bow. It was a powerful instrument and the arrows were long and barbed; these were used in such enormous numbers that special supply trains of camels were used to transport them. The Parthians gave a convincing demonstration of the fact that tactics are an essential concomitant of firepower. Although they were an organized army they used the tactics normally associated with outnumbered guerrilla forces. They occasionally ambushed their opponents, but were usually content to draw them on while wearing them down with harassing tactics. Endlessly circling the advancing columns, they tormented and weakened their enemies by constant showers of high-flighted arrows. If the Romans sent out a cavalry squadron to bring them to battle, the Parthians galloped away: just before they disappeared they would wheel round and discharge a final flight of arrows – the renowned Parthian shot. Eventually, perhaps, when they deemed that their opponents were ripe for defeat, the Parthians would launch their complete army in a devastating surprise attack. If swift victory ensued, that was the end of the story, but if not the Parthians would withdraw and once more begin their circling and harassing tactics.

The variety and sophistication of warfare in the centuries immediately preceding the birth of Christ may make the modern soldier feel glad that he does not have to pit his personal skills against such opponents – armed with weapons equal in power to his own. As Field-Marshal Montgomery frequently reiterated in the Second World War, an unfit soldier is a useless soldier. During and after the war the achievements of the SAS have been built on an exceptionally high standard of personal fitness and endurance, but even the SAS would find something to envy in the skills of people like the Numidians, who came from what is now Algeria. They had nothing in their favour: they were small in stature and they had small horses;

they had no armour, no saddles, no reins, no bridles, and nothing but javelins with which to kill their opponents. Yet despite every conceivable disadvantage they were the terror of the armies of their day. The Romans, who suffered much from them, wondered whether these semi-naked warriors could be human. There was no terrain so rough and dangerous that it could slow them down, let alone impede them. Like the Parthians, they excelled in harassing tactics. Added to their skills as cavalrymen was that other great asset to firepower – the reputation of invincibility. In the course of the book it will be encountered on many occasions.

The bow, and arms derived from it, were a constant weapon throughout history. The vast numbers of flint arrowheads which have been discovered on long-forgotten battlefields in many different countries show how widespread its use was. Sometimes other weapons have been more popular, but never for long. It was eventually superseded by firearms, but not until some hundred years after their introduction were firearms actually superior to the bow. In time the bow went out of fashion as a weapon of war because the larger numbers involved in later wars made the task of training sufficient bowmen impossible. The devastating longbow of the Middle Ages required a man to practise several hours every day until he was proficient. The recruits pressed or cajoled into the army in the eighteenth and nineteenth centuries would not have had the physique for the task: they were the sweepings of the towns, whereas the medieval bowmen had come of sturdy yeoman stock.

That the bow could be a battle-turning weapon was shown at Hastings in 1066. In the Bayeux Tapestry the bows shown are somewhat smaller than they would have been at the time: the tapestry is artistic rather than accurate to the last detail. Hastings – which was not actually fought at Hastings, but at Senlac, nearly 7 miles away – was a battle in which the Norman knights had tried all day long to charge up the slopes and destroy the élite Saxon housecarles. Those who reached the circle of axes were flung back by slashing blows which bit through shield and armour. Twice the Normans tried to lure the Saxons down from the hilltop by feigning retreat, but this deceived only the less disciplined of their opponents. Finally, when a fresh supply of arrows had arrived for the Norman bowmen, they tried flighting them high in the air so that they would fall on the Saxon centre. This manoeuvre, combined with a final vigorous charge, brought the Normans victory. It was a form of tactics which rarely failed. The Parthians had used it, and it would be employed again and again, even when the dive-bomber and helicopter gunship replaced the high-flighted arrow. Similarly the technique of underground mining and frontal assault, used throughout the Middle Ages, never varied: the materials were different but the principles were the same. Firepower was made up of a combination of weapons and tactics, not least of which was an assault on morale (see Chapter 12).

Another embellishment of the arrow was to barb the head. Arrow wounds were said to be excruciatingly painful, but the life of a wounded man could sometimes be saved if an arrow was pushed right through his body; the head was then cut off and the shaft withdrawn. In order to prevent this happening, some arrowheads were fixed loosely on purpose, so that the head would be left in the wound. The practice of pushing an arrow through was prevented by fitting it with barbs facing both ways.

The efficacy of arrows was frequently enhanced with poison, as described earlier. Poisons are mentioned as being in use as early as biblical times: the Parthians and Scythians certainly had them. Everyone complained bitterly about these treacherous and cruel devices, but everyone used them. Poison was not, of course, limited to arrowheads: poisoned swords are mentioned everywhere from the ancient Norse sagas to medieval Japan. Bamboo spears and darts were particularly suitable for carrying poison; the point of the assassin's dagger was often coated with it, and many an unlucky victim who thought he had escaped with a mere scratch had an unpleasant surprise coming to him. More surprisingly, the caltrop often carried poison. The caltrop, a word said to be derived from Latin *calx* = heel and *trappa* = trap, became all too well known in medieval times when it took the form of four sharp spikes so joined that one was always pointing upwards. They were scattered profusely on ground over which horses were expected to advance, but they could be almost as deadly against foot soldiers, whose crude shoes gave virtually no protection. The caltrop has been revived and used extensively in the twentieth century as a tyre-burster. In the Second World War it was used in the Western Desert and by saboteurs in France. In a different form it has been used for stopping suspect vehicles. No one who tries bursting through a roadblock continues very far or fast if he has four shredded tyres.

Probably the greatest triumph of the unpoisoned bow in a demonstration of concentrated firepower occurred at Agincourt in 1415. The weapon in question was the English longbow which, in spite of its name, was probably used mainly by Welshmen or Scots. Although the power of the longbow had been known since pre-Christian times, it did not pass into general use until the Welsh demonstrated its potential against the Normans in the eleventh and twelfth centuries. An example of its power comes from the Welsh chronicler Giraldus Cambrensis, writing at the end of the twelfth century: one passage mentions arrows which penetrated 'four fingers thick' into the oak door of Abergavenny Castle. An even more dramatic example was verified by William de Braose. When he was leading a small invasion force into Wales one of his soldiers was wounded by an arrow which, dropping out of the sky, cut through the chain mail on his thigh, went right through his leg and then penetrated the horse. As the horrified soldier turned to ride out of battle a second arrow (doubtless one of many) pinned him through the other leg in precisely the same way.

The longbow, as it came to be known, was not a simple weapon, even though it may look it. It was 5–6 feet in length and made of yew, which had the virtue of being able to stretch on one side while being compressed on the other, if it was cut properly from the parent tree. After being bent almost double, the bow would spring back to its original shape and retain it. The only drawback to this formidable weapon was that it required considerable strength and skill in the user. Bowmen were therefore required to practise regularly and to take part in numerous competitions. The end of a session of archery practice was proclaimed by sending a single arrow high into the air where it would be seen by all. This was known as the upshot, a term which eventually came to be used as the summary of an argument or discussion. As well as practising with his bow, the archer had to look after it and protect it from blows and cuts which might spoil its accuracy. Accuracy in aim was important; at 100 yards an archer was expected to be able to hit a moving target, and 240 yards was well within his compass. Not many archers would have cared to try to split an apple placed on their son's head, as William Tell is reputed to have done, but there were, no doubt, a number who were capable of such a feat.

The bowstring was attached to notches at each end of the bow, from which, as mentioned earlier, it could easily be removed if there was a danger of its being wetted by rain. It was frequently greased to protect it and keep it flexible, and the medieval archer would need to carry a small quantity of the appropriate mixture with him. He also carried an arm guard to protect his wrist, a shooting glove and additional finger stalls. He wore a belt on which a quiver was suspended, and one of his first actions on going into battle was to withdraw a dozen arrows from the quiver and place them on the ground by his left foot. Once the order to fire had been given, he would launch off the first arrow and bob down to pick up the second. In this way he could put twelve arrows in the air in a minute, and the appearance of a company of archers would resemble a group of ballet dancers practising their exercises. It was this ability to produce devastating concentrated firepower which made the English archers so dreaded in the Middle Ages. At Agincourt five thousand archers, each launching twelve arrows a minute, would create a barrage which would first blacken the sky, then descend with crushing force on men and horses. The weight of a clothyard shaft, as the 3-foot-long arrow came to be known, could drive the greased bodkin points through chain mail and leather jerkins. Horses were the principal victims, for it was impossible to protect them properly. Once unhorsed, their riders were too much encumbered by their heavy armour to fight effectively, even if they could stand upright – as was almost impossible on the greasy slopes of Agincourt.

The average length of an arrow was 33 ins, and they were made of whatever suitable wood was locally procurable. Thus in Europe hardwood was used, while in the Far East bamboo was more easily available. Feathering was not universally practised, and both feathered and unfeathered arrows are recorded from the earliest times. 'Feathers' were sometimes not feathers at all but a different substance such as a stiff leaf or sliver of wood. A carefully placed feather could make an arrow turn in the air and reach its target while spinning like a bullet from a rifle. Medieval English arrows were as simple as they were deadly. They were usually fletched (fitted) with three goose feathers. Geese were plentiful and their feathers were, fortunately, very suitable. Medieval accounts contain orders for huge quantities of feathers, shafts, bows and bowstrings (in England usually made of linen or hemp). Medieval logistics required substantial stockpiling. Richard I took fifty thousand spare horseshoes when he set off on the Third Crusade. Castles such as Northampton and Warwick stored arrows by the thousand; they were supplied in sheaves of twenty-four. At the siege of Bedford Castle in 1215 it is known that there were nineteen thousand crossbow quarrels, of which all but nine hundred were used, and it is reasonable to assume that there were at least twice that number of longbow arrows. Siege warfare was a very serious business and could hold an army in position for up to a year. Maintaining morale among the besiegers was important, so some of the arrows were expended in competition shooting or hunting in the nearest forest.

It seemed to most people that, as the use of guns was perfected, the day of the bow was over, even though it was probably more accurate than many of the guns used in the Crimean War of 1854–6. But the Second World War, with its collection of unorthodox regiments, revived interest in this apparently obsolete weapon. It could be manufactured easily with modern materials, it was silent, it was accurate, and it was thus an ideal weapon for a raider to use when picking off an enemy sentry. A very strong supporter of the longbow as a modern military weapon was the Norwegian Anders Lassen, VC, who had come to England at the outbreak of war and joined the Commandos; later he transferred to the Special Air Service (SAS). He was a born hunter, of whom it was said that he was capable of stalking a stag and killing it with a knife. Lassen was so convinced of the value of the bow that he wrote a letter to the War Office on the subject, listing the weapon's advantages very persuasively; the War Office response was not to equip whole companies of archers, but merely to send Lassen two bows and arrows. There is, however, no record of his successfully using them in battle; when at close quarters with the Germans he seemed to prefer grenades and sub-machine guns.

2

The Birth of Artillery and Missiles

In the first recorded battles firepower was related to numbers and hand-held weapons. The more soldiers – whether bowmen, spearsmen or swordsmen – you could put into the attack, the greater your chances of victory. Within those limits, training and tactics were investigated very thoroughly by the early battlefield commanders. In ancient China Sun Tzu, who wrote a book entitled *The Art of War* in 500 BC, laid down maxims which are as true today as when first written: one of the most important of them was that large numbers are but lambs led to the slaughter unless they are properly trained. Most of his precepts are devoted to increasing the effective firepower of one's own force, but with the exception of horses and chariots this had to be accomplished by manipulating men rather than materials. They must be motivated to give of their best, and commanded to attack at the appropriate moment ('The victorious strategist seeks battle after the victory has been won, whereas he who is destined to defeat first fights and afterwards looks for victory.') He stresses the importance of strategy, tactics and manoeuvre and, above all, emphasizes the value of surprise and deception.

However well trained one of Sun Tzu's armies might have been, it would have been virtually useless against a well-fortified town if that town was properly defended. Sun Tzu did not approve of siege warfare, as it took up too much time, and he would have disapproved of it even more if he had been compelled to fight in the Middle East rather than in China. Fortified cities existed in the Middle East some two thousand years before Sun Tzu was born; the earliest-known one was Ur of the Chaldees, which had walls 23 feet thick. It also had towers at intervals along the walls, and a moat outside. A fortress of that calibre could not be captured by manpower, unless that manpower was exceptionally resourceful and the defenders correspondingly negligent: it needed to be breached by engines. The first siege engines were used by the Assyrian Army, which had both missile launchers and armour-protected vehicles to bring them into close range. We know little of this early artillery in the Assyrian forces, but it was clearly well established even at that distant time. Artillery would later acquire different and varied methods of propelling its missiles, but the same principle would govern its use until it met its greatest test of all, the destruction of the defences on the Normandy beaches on D-Day in 1944. Much of that later destruction was accomplished by aircraft and long-range naval guns, but the general principle remained the same: if you want foot soldiers to penetrate enemy defences you must first cut a way for them through the main obstacles.

Artillery weapons which were in use in 500 BC were being used with little variation nearly two thousand years later. Like all later versions, they were of two main types: they either fired directly at the target with a low trajectory, or they launched their missiles with a high trajectory, expecting them to fall on the far side of any defences. In the first category came the ballista, which propelled a giant, iron-headed arrow at weak spots such as town gates; from this are descended many modern weapons, including anti-tank guns. The second type included the catapult, which had a variety of intriguing names; it slung its missiles high into the air, thereby demonstrating the future possibilities of howitzers and trench mortars.

The power of this early form of artillery should not be underestimated. The ballista derived its name from the Greek word *ballo* = to throw, but in fact launched its missile directly at its target by the tensile power of a giant crossbow. This fearsome weapon dug itself into the stonework of the defence and weakened the whole structure. Neither the ballista, nor its lethal sister, the mangonel, was used in isolation. A besieged fortress would be surrounded by dozens of ballistas and mangonels, launching their missiles in sequence or in salvoes. A medieval fortress under siege, with someone trying to set light to the door, someone else approaching its battlements with assault towers, someone crashing a battering ram into the lower walls, someone mining underneath, all to the accompaniment of a constant rain of arrows and great missiles from siege engines, was no place for the weak or irresolute. The mangonel, which gave us our word 'gun', was a torsion weapon. It consisted of two stout posts, usually mounted on a mobile platform, with two elastic ropes between them. The ropes were occasionally made of horsehair but were preferably of human hair, which is extremely tensile. A few years ago a mangonel was made for a television programme but, as human hair of the right quality and length was not available, nylon was used instead. It proved an unsatisfactory substitute.

A beam was placed between the ropes, which were then twisted until sufficient tensile power had developed. A large stone, perhaps weighing 100 lb or more, was put in the cup or sling at the end. The beam was then released. This might appear to be a somewhat haphazard process, but there are numerous reports of mangonels being able to launch their missiles with amazing accuracy as to both distance and target. On the bigger mangonels the missile holder seems to have been large enough to accommodate the major portion of a dead and much diseased horse. It was thought that this unwelcome present would not merely spread disease but also make the area so foul that an early surrender would seem appropriate. Captured spies and suspected traitors were also placed in the launcher, sometimes with no hope of reprieve, at other times with the intention of encouraging them to disclose useful information. The medieval world was infested with spies; it seems that our present practitioners of the art follow a long, if not exactly honourable, tradition.

The principal missile launchers used in the Middle Ages, the ballista and mangonel, had both been invented over a thousand years before. However, the medieval engineers did make one contribution – the trebuchet, a mangonel which worked by counterpoise. The launching lever was balanced between two uprights. To one end a heavy weight was attached; at the other was the cup containing the missile. Once the missile had been placed in position this end of the beam was hauled down by means of ropes and impressed labour; this process pushed the weighted end high in the air. When the order was given, the ropes were released, the weight descended with the force of gravity and the missile curved through the air towards its target. The trebuchet was popular because it could be transported in sections, dispensed with the tensile hair and winding mechanism, and could be weighted with a convenient piece of rock.

But then came gunpowder and the day of the catapult was over – until it dawned once more as a means of launching aircraft from carriers in the mid-twentieth century. The inventor of gunpowder is not known; it was probably discovered accidentally by several different nations. The Chinese were early in the field, but used gunpowder for fireworks only. When it appeared in the Middle East in the early fourteenth century it was used to project an arrow from a bamboo gun. Gunpowder is extremely simple to make, consisting merely of saltpetre, sulphur and charcoal. It burns quickly, producing a cloud of white gas, and it is this gas which projects missiles when the powder has been burnt in a suitable container. Surprisingly it was not superseded by other explosives until the mid-nineteenth century, when nitroglycerine and nitrocellulose took its place.

Gunpowder was first mentioned in writing by Roger Bacon, an English friar, in 1249. The earliest picture of gunpowder being used in a weapon is found in the Millimete Manuscript of 1327, which is in the library of Christ Church, Oxford. The Millimete 'gun' is a vase-shaped container with a feathered bolt protruding from it. A knight is shown setting light to the fuse. The feathers appear somewhat unnecessary, and the container was doubtless much too large and therefore overcharged. However, the vase shape remained popular for a while; the Italians called their firearms *vasi*. This new discovery must have been more unpopular with knights than crossbows had ever been, but at least it had the advantage of being expensive, and therefore a suitable weapon for a king to monopolize. Cannons were said to have been used at Crécy in 1346, but the evidence is tenuous and, if they were used, they seem to have had little effect on the battle – which was won by the longbow. Cannons, mortars and bombards, mostly firing stone balls, were used at the siege of Orleans in 1428, but a more inspired and successful use seems to have been made of them by the Hussites in Hungary in the same period. This success was all the more remarkable because they were employed by a peasant army. It was commanded by a military genius, Jan Zizka, who mounted his cannons in four-wheeled carts and gave the infantry cover with them. Zizka also employed pikemen and crossbowmen, with cavalry waiting poised in the wings. Not surprisingly, he won a continuous series of victories against all the knights sent to suppress the Hussite revolt.

For an invention of such obvious importance there is remarkably little information in medieval chronicles. This may, of course, be due to the fact that the chroniclers, usually monks, realized how unpopular this new development must have been to the armoured knights, who were often their patrons. But there was soon a quick transition from underestimating to overestimating the potential of the new weapon. This phenomenon is not uncommon with new weapons – gas is another example. In 1464 the otherwise impregnable castle of Bamburgh, in Northumberland, was quickly beaten into submission more by the threat of gunfire than by its effects, and neighbouring strongholds also belied their reputations for the same reason. But perhaps the most surprising fact about the early firearms was the imagination shown in their use – and the neglect which ensued. In the mid-fourteenth century a weapon called a ribaudequin appeared; it consisted of a number of tubes bound together so that all could be fired simultaneously or even made to revolve while delivering their projectiles. Three which were used at Verona in 1387 consisted of 144 tubes each and fired salvoes of twelve rounds. Obvious drawbacks were that they were heavy to transport and difficult to handle, one burst tube could destroy the rest, and the need for such a concentration of fire was not always evident. Not until the Second World War was a proper application of this design seen, in the dreaded German Nebelwerfer, described in Chapter 8.

Many of the setbacks encountered in later weapons were equally well known to those handling their predecessors. Not least was the dud – welcome enough to the recipient, but exasperating to the firer. An astonishing number of duds were fired in both world wars; sometimes their failure to explode was caused by deliberate sabotage but more often it was the result of mechanical breakdown. Rather more disturbing was the tendency to explode in transit, causing unbelievable havoc on ammunition trains. Today, over forty years after the end of the Second World War, huge bombs weighing up to 4000 lb are still discovered at intervals: they were dropped in the period between 1940 and 1944, but for a variety of reasons buried themselves in the ground and waited to be discovered by an unlucky excavator. Some of them had been equipped with percussion fuses which had obstinately refused to function.

One of the earliest and most distinguished victims of an artillery accident was King James II of Scotland in 1460. He was at the siege of Roxburgh Castle, standing near a large bombard made of semi-circular plates bound with hoops. It had been specially brought over from Flanders, where it had been very successful and earned the nickname 'The Lion'. Unfortunately for James, it exploded and killed both him and many of his followers.

Early gunpowder was unreliable and unpredictable. It was rammed down the barrel of the cannon and then ignited through a touch-hole in the breach. Too small a charge produced an ineffective, short-range shot; too much was likely to burst the cannon and kill the gynours, as the crew were called. Whether the word 'gynour' has given us the word 'gunner' or 'engineer' is open to speculation.

Several disastrous experiences demonstrated that lighting the gunpowder through the touch-hole was not the best method and it was therefore replaced by a train of quick-burning powder. Having lit the end nearest to him, the firer then retreated rapidly to a safe distance. If, however, the powder train sputtered out, he had to come much closer and reignite it where the break had occurred – an unpopular but necessary task. Early chemists seemed slow to learn, even by trial and error. Although Roger Bacon had emphasized that gunpowder should be dampened (but not too much) to prevent the ingredients separating themselves into non-explosive layers, his suggestion was often ignored; the key to an effective explosion was to have the ingredients ignited when the mixture was in exactly the right proportions.

The main hazard to early artillerymen, apart from the gun itself exploding, came from fumes and the blowback. The fumes were no doubt a considerable health hazard, but the dangers from the recoil exceeded this. Sometimes the back of the gun blew out while leaving the projectile unmoved; at others a stream of burning gunpowder would spray out of the touch-hole, burning or blinding everyone within range. Throughout history, soldiers have come forward readily to handle new weapons, however complicated or dangerous, if they possess impressive firepower. In doing so they know they are making themselves into corps d'élite, and they become very jealous of their preserves and privileges. The respect given to the firepower of the weapons they handle is soon transferred to the user. In modern armies the opinion is frequently voiced that the gunners have never rid themselves of these estimates of their own importance – a view dismissed by the gunners as mere jealousy.

Artillery soon divided into two main types: the gun and the bombard. The gun was the descendant of the ballista and fired straight, or with a very low trajectory. This category included handguns of various types. The bombard was the successor to the mangonel and usually consisted of a broad, bucket-shaped instrument cast from brass or copper. It was dragged around on wooden sledges and elevated or lowered by using a system of wedges. Bombards, even more than guns, were liable to scatter burning powder around them as they launched their missiles.

Both types of gun fired solid shot. For many years stone cannon balls were popular, as they were easy to collect and shape. Visitors to castles will often notice piles of stone cannon balls which have either been dug out of the ground or retrieved from the moat; some may have even performed as ammunition for mangonels. On occasion, stone was replaced by solid iron shot. Some rather unsuccessful experiments were made with hollow iron cannon balls which were then filled with gunpowder. This type of missile was only effective if the gunpowder could be ignited before it sped on its way. The task of lighting the longer fuse for the missile immediately before setting light to the shorter fuse for the gun was a delicate and frequently unsuccessful operation. The round shot were named grenades, and the oblong became bombs. A more successful form of grenade was the smaller version which was later issued to individual soldiers. These were held in the hand, lighted by means of a fuse, and then thrown in the direction of the enemy. As it was important to know in which direction to throw, tall men were favoured for the regiment of grenadiers which later came into being. It is not mere coincidence that the Grenadier Guards stipulate a minimum height of 5 feet 8 ins for their recruits.

Inventors, of varying degrees of scientific skill, were not lacking in the early days of artillery; but then, as now, there seemed to be a considerable gap between what was feasible on the drawing board and what would work in practice. Red-hot shot figured in one of the experiments: the process of placing red-hot metal on to a gunpowder charge might make anyone thoughtful. More successful tests were carried out with hollow cases containing lead shot. One of the more ambitious of the early experiments may now be seen at Edinburgh Castle. It is a fifteenth-century iron bombard, named Mons Meg, which weighs 5 tons, has a bore of 30 ins and fired an iron or stone ball of 300 lb. One suspects that the range of this and similar monsters was small (although one mile was claimed for Mons Meg); but their presence was intimidating or encouraging, depending on which way they were facing.

In general more imagination and ingenuity was shown in the manufacture of ammunition than in the guns to propel it. Case shot (shot which scattered) and chain shot were early on the scene. The main problem with guns lay with the material used in their manufacture. An early form of steel (first used at the beginning of the thirteenth century) was smelted by heating iron over charcoal, but any impurities left in the metal were likely to cause it to split at inconvenient moments. Steel is an alloy of iron and small quantities of carbon which can be tempered to various degrees of hardness, and is less brittle and more malleable than cast iron; it was therefore altogether more suitable for fashioning into gun barrels than anything which had preceded it.

Medieval cannons with a long barrel soon acquired the name of culverin, derived from the Old French word *coulevrine* = snake. Their long, tubular appearance bore a rough resemblance to a snake, a feature enhanced by the custom of carving a snake's head around the muzzle. A full culverin fired a 16-lb shot far enough to make it a very useful weapon for naval warfare. Smaller-size guns were known as demi-culverins. These terms were much in evidence in the English Civil War of 1642–5, but became obsolete when guns were classified by the weight of the projectile they fired, for example 18 lb. During the fifteenth century gunnery became less a matter of chance than of science. Different nations produced their special contributions, and the Italians were quickly to the fore in the field of ballistics. At that period military developments tended to be freely shared rather than jealously guarded, as nowadays, and there was no ignominy in working for the monarch of a foreign country. Henry VIII, King of England 1509–47, employed a Fleming by the name of Hans Poppenruyter to supervise the manufacture of his artillery, and even established a school of instruction for English artillery artificers. Poppenruyter was responsible for the provision of 150 guns, among them twelve formidable bombards known as the Twelve Apostles.

There was, of course, every reason for England to respect the abilities of artillery, even at this early date. Although the spectacular English victories at Crécy, Poitiers and Agincourt, some two hundred years before Henry's death, are well known to English schoolchildren through the enthusiastic descriptions of them in history textbooks, the battles by which England eventually lost its foothold in France have been less fully reported. The main reason for England's discomfiture was Joan of Arc – not because, mounted on a white horse, she led triumphant armies of medieval warriors, but because she established the point that women might offer common-sense views about military tactics. Her successor as an influence on French policy was Agnès Sorel, mistress of Charles VII, King of France from 1440 to 1450, and she persuaded the King that unless he established a standing army which was properly paid and equipped his country would never escape the humiliation of an English army of occupation. Once that army was established, the need to equip it with the latest weaponry was apparent. The French military scientists therefore developed the culverin so that its dangerous recoil was absorbed; it was made mobile by being mounted on wheels rather than on a sledge, and its elevation was accomplished by a screw mechanism instead of wooden chocks. Equipped with these weapons, the French were able to beat an English army of slightly superior numbers at Formigny in 1450. There were, of course, many reasons for the English defeat, not least being the fact that the army of 1450 bore little resemblance to the Agincourt victors of 1415, and the English monarchs of each period, respectively Henry VI and Henry V, were in no way comparable in military leadership. Nevertheless the main point proved at Formigny was that a well-trained army, equipped with the latest technology, will always defeat equal or even superior numbers using inferior equipment, *provided* the correct tactics are used.

Agincourt had been won by the English because the French had tried to advance in a solid mass and the English archers had therefore been able to destroy them from long range. At Formigny the French were able to harass the English bowmen with culverins deployed on the wings. Goaded to desperation by this fire, to which they could make no effective reply, the English longbowmen came out of their defensive position and advanced over open ground as infantry. Armed only with weapons quite unsuited to such tactics – for which pikes would have been ideal – they fell easy victims to French cavalry charges.

Two years later the decisive influence which artillery would have on future battlefields was once more demonstrated, this time at Castillon. An army led by the charismatic John Talbot, Earl of Shrewsbury, landed at Bordeaux with the intention of relieving the besieged town of Castillon and restoring the military situation in France. But by this time the French had increased their artillery strength dramatically and were able to confront Talbot with a hundred guns. His attempt to charge through the middle of the besiegers simply meant that their guns were turned on him and his army was cut to pieces. Talbot died fighting gallantly: his body could only be identified by a tooth, but courage, as would be proved over and over again in the future, was of no avail against sheer weight of metal.

These were examples of the influence of gunpowder on the battlefield, but an even more striking demonstration of its power was shown at Constantinople in 1453, the year after Castillon. The city, with its high, heavily fortified walls, was thought to be impregnable, but this view was not held by Mohammed II, the Ottoman Turkish commander, who approached its perimeter on 5 April that year. He had a formidable array of bombards, as well as other guns, which had been made by a Hungarian with the misleading name of Urban. Mohammed's bombards threw stone shot weighing up to 1500 lb, and as he had fourteen batteries, each containing thirteen of them, as well as a considerable number of smaller cannon, it was obvious that the walls of Constantinople would be lucky to survive this onslaught. In the event, they did not, and seven weeks after the assault had begun there was a breach wide enough to admit Mohammed's troops. It was the end of an era, and not merely a military one.

Until 1512 artillery battles in the field had tended to be won by the best equipped, which usually meant those possessing artillery. However, the time was bound to come when two armies with well-matched equipment would meet each other: it occurred at Ravenna, when an army of the Holy League (the Pope, Spain, Venice and Switzerland) tried to eject the French from Italy. A mutual bombardment was conducted with skill on both sides, but eventually the Spaniards were forced out of their defensive position. Once they were in the open they were obviously vulnerable to a French charge combined with an encircling movement. In the last stage of the fighting the French commander, Gaston de Foix, was killed in a hand-to-hand fight. Though only twenty-three he had already established a reputation as one of the most daring and skilled commanders of the age.

Artillery changed men's entire attitude to warfare, not merely on the battlefield but in the industrial and logistical requirements of armies as well. Although the longbow had demonstrated that an enemy could be defeated while still so far away that he was scarcely visible, the additional concept that an army or fortification could be pounded to pieces, if cannon firepower was powerful enough, was only now being appreciated. However, the material performance of artillery still fell some way short of its psychological effect, in which not the least of its assets was the intimidating noise it made. In addition, there was always the haunting fear that the opposition might have some new form of cannon or ammunition, of greater power than any known hitherto, and was therefore now invincible. All over Europe faith in fortresses and walls was disappearing much faster than stonework disintegrated under shellfire. Once more, men looked to the protection of the earthen burrow, rather than anything above it. Many of the old stone castles, formidable structures with walls up to 20 feet thick, were prematurely abandoned for trenches and earthwork bastions. A whole new concept of fortification was devised, a concept in which surfaces were made to deflect cannon balls, while at the same time giving the defender an opportunity for flanking fire. Military architects no longer concerned themselves with fortifying residences, such as castles or manors; instead they examined methods of protecting towns by those complex seventeenth-century layouts for which the Frenchman Vauban is especially famed. These fortifications, a combination of stonework and earth, may still be seen in many old European cities, such as Vienna, Antwerp, Ostend and Sedan. Words such as bastion, ravelin, caponier, scarp and glacis were now freely used in military planning. Some of these defences became so complex that the task of commanding the garrison inside was made almost impossible. Vauban was, in fact, only one of the architects who achieved fame in the field of fortifications: others were Cormontaigne, Montalembert and Carnot, who did not hesitate to criticize the designs of the great master himself.

Military architects did not, however, confine their attentions entirely to towns. Although castles were now considered to be death traps rather than refuges, some were so strategically positioned at nodal points that efforts had to be made to protect them. This protection usually took the form of outworks, which also acquired a vocabulary of esoteric terms. Thus a sconce was a square fortification with elaborate corners, a redoubt a forward defence, and demi-lunes, hornwork or redans were small fortresses which had to be reduced before the main building could be attacked; they were all examples of the principle of trying to cripple the enemy before he came too close. Such elaborations no doubt both gave confidence to the defender and presented some problems to the attacker, but there was reason to believe that these developments – and the mystique attached to the terminology – were often of less value to the soldiery than to the reputations of their designers. The English Civil War demonstrated that simple medieval flint rubble castles could put up an effective resistance against the best that the latest weapons could hurl against them. This does not seem surprising when one remembers that they were constructed to stand up to bombardments of stones weighing 100 lb and over: few seventeenth-century cannon balls could match that firepower. Castles such as Denbigh in North Wales and Corfe in Dorset held out for periods of over a year against everything the artillery of that age could hurl at them.

More surprising, perhaps, are two episodes from more recent military history. During the Second World War the ancient walls of the city of Bordeaux stood up to a day's pounding by German artillery; even more impressive was the impregnability of the centuries-old Fort Dufferin in Mandalay, Burma. In March 1945 the retreating Japanese had taken refuge in this fort, which was protected by a moat and crenellated ramparts. The commander of the British 19th Division, General Rees, battered the walls with a bombardment from breaching guns. As they made remarkably little impression, aircraft were brought in to add their quota to the pounding. Among other methods tried was that of bouncing bombs along the surface of the moat to penetrate what was thought to be the weakest point in the defences. After five days, a direct frontal attack was made by infantry. All to no avail. The bombardment began again with doubled vigour. Breaches now began to appear in the walls and a further infantry assault was contemplated. Fortunately, perhaps, the Japanese had decided it was time to leave, and what would have been a costly assault operation was avoided. But this had been a striking example of the fact that medieval-style fortifications, long thought to be obsolete, could put up a spirited resistance to the best of modern firepower.

However, in the early days of artillery lack of confidence in buildings was soon replaced by over-confidence in trenches. Masonry and later concrete had demonstrated that, although some bullets and cannon fire could be checked above ground, the best defence apparently was to dig in. But even below ground the soldier was vulnerable. A direct hit could seal him into his refuge, even if it did not kill him outright. Living below ground level was bad for the health and even worse for morale. Major-General G. Hopkinson, founder of the 'Phantom' reconnaissance regiment in 1940, refused to allow men to take refuge in slit trenches for, he said, 'Slit trenches make cowards.' A legacy of the Second World War was the term 'Maginot mentality', which, after the Maginot Line had been penetrated by the Germans, indicated that if soldiers are allowed to live in supposedly invulnerable fortifications they soon lose all their combatant spirit.

In the twentieth century these two former methods of withstanding firepower have been combined. The First World War saw the introduction of deep dugouts as a part of trench warfare. The appalling British losses at the Battle of the Somme in 1916 were partly caused by the fact that the Allied Higher Command had no conception of how deep and secure the Germans had made their dugouts. The British view on 30 June 1916 was that concentrated artillery barrages had destroyed all the German fortifications at ground level, and had also cut great gaps in the barbed wire defences. But when the attack went in, on 1 July, the British infantry was mown down in swathes. The bombardment had failed to cut the necessary holes in the barbed wire and the Germans, far from being annihilated in the preliminary bombardment, had simply retired to their deep dugouts, where the only inconvenience was foul air. As the British attacks came in, German infantry poured out of these secure strongholds and manned machine guns behind the uncut barbed wire. The result was forty thousand British casualties within the first three hours.

But this lesson of the value of the deep dugout, learnt at such appalling cost, was not forgotten. Although in the Second World War the majority of the British population was provided with – and had to rely on – surface shelters known as Andersons, vital installations and personnel were protected in deep subterranean caverns. One in St James's Park in London, which has only recently been opened to the public, contains rooms in which Churchill and his staff could continue to work even if most of London was in ruins. There were similar deep refuges elsewhere.

After the Second World War, when firepower had been increased to horrendous proportions by the atomic bomb, the number of deep shelters increased in all countries which felt vulnerable and could afford them. Probably the most impressive is in Nebraska at the headquarters of the US Strategic Command; this is almost an underground city. Doubtless the Soviets have similar, though not perhaps so elaborate, installations. All European countries took similar precautions, and deep shelters would house not merely wartime headquarters but also planning groups for local civil defence. The general public, although well aware of their existence, has been kept in ignorance of the exact location of these hideaways. The attitude towards them tends to be philosophical: if nuclear war ever comes, it would devastate the countryside to such a degree that even those who survived the initial bombing might subsequently come to wish that they had not, so little would there be to live for. However, this pessimistic view is not shared by the military planners, whose experience in the pre-nuclear age has encouraged them to believe that, however great the seeming destruction, much tends to escape and survive. The answer to absolute firepower, which is a possible definition of modern nuclear warfare, is to build below the ground what used to be above it: the castle and the fortress would therefore be underground, surrounded by the civilian population. This may appear a fanciful concept, but is no more so than colonizing the moon or another planet, or even establishing cities on space stations in orbit. At the time of writing there are some fifty thousand nuclear weapons in existence and, if even a small fraction of that number is ever used, the only hope for mankind will lie well above ground level or well below it.

To those who hope to survive a nuclear holocaust deep shelters and subterranean cities give great comfort. Others may feel it is better to be doomed than entombed.

3

Small but Deadly: the Personal Weapon

Small arms, in other words firearms which could be carried and fired by the individual on the battlefield, were first seen at Perugia in Italy in 1364. They were in fact more akin to modern mortars than to rifles or pistols, but these undoubtedly were the first hand-held weapons to use gunpowder. Contemporary manuscripts show them to have been miniature bombards mounted on staves some 7–10 feet long. For firing they were propped on a tripod which allowed the butt to rest on the ground and take the recoil, while the bombard, muzzle-loaded, projected a small missile at an angle of about 45 degrees. They must have been difficult to operate, for they were fired by applying to the touch-hole a continuously smouldering cord known as a match, and, as everyone knows who has ever tried to light a fire in the open when there is even the slightest breeze, the chances of success are low. In the following century the stave is shown to have become much shorter, enabling it to be held to the shoulder and aimed at a target. This, however, must have made firing an even more delicate process.

Needless to say, the weapon had no immediate appeal to the English with their rapid-firing, reliable longbows, and not until the disasters of Formigny and Castillon did they consider it as a serious weapon of war. Consequently, by the time the English did adopt the handgun it had progressed a long way from the primitive weapon used at Perugia nearly two hundred years before. It had also become known as the hackbut in Germany, and the arquebus in France; the English used both words. The arquebus, which bore some slight resemblance in shape to the modern rifle, was fired from a pan which projected from the side of the barrel. The touch powder was ignited by a match, and from this lighted touch a flame licked back into the barrel, which held a powder charge behind an iron ball. Two great weaknesses of the weapon were the ease with which the touch powder could blow out of the pan, and the almost equal ease with which the match could go out at the critical moment. A sudden shower of rain could have as diasastrous an effect on arquebuses as it could ever have had on crossbows at their most vulnerable. However, the most serious contemporary criticism of the arquebus was that it lacked range, so it was made larger and the barrel was lengthened. These changes made it too heavy to hold in the aiming position, and a forked stick was required to make a rest for the barrel. Although cumbersome, the altered weapon, now called the musket, was sufficiently effective to be retained for nearly a hundred years. As with the artillerymen mentioned in Chapter 2, the fact that the new weapon was complicated to handle and fire accurately gave its users, the musketeers, a certain prestige.

A further notable disadvantage of the musket was its slow rate of fire, which rarely exceeded one shot every two minutes; it was said that twenty-four separate movements were required between firing one shot and being ready for the next. Nevertheless it proved its worth at the Battle of Pavia in 1525. King Francis I of France was besieging the city with a Franco–Swiss force numbering 28,000 when he found himself confronted with a 2300-strong mixed German, Spanish and Italian force representing the Holy Roman Emperor, Charles V. Although the French fought with considerable gallantry and skill, the Spanish musketeers poured in sufficient deadly volleys to disrupt both the cavalry and the gunners. This French defeat showed that the cavalry was no longer the master of the battlefield and that, in future, infantry armed with ball ammunition would play a vital role in securing victories.

Inevitably, experiments were carried out to develop an even smaller weapon. The call for this came from German cavalrymen who could see the merits of their hackbuts but found them too cumbersome to use when on horseback. But before the right weapon could be produced, a more efficient firing mechanism was required.

Initially this was the wheel-lock, whose mechanism was crude but effective. A toothed wheel alongside the barrel was wound up on a spring. When the soldier pulled the trigger the wheel, driven by the spring, revolved rapidly and hit its teeth against a piece of flint or other suitable stone fixed near the pan. The spark then ignited the powder which sent the bullet on its way – and it could all be done with one hand. Pistols such as this were good enough to spell the doom of pikemen, who had long been dreaded by cavalry – they could easily stop a cavalry charge by taking up a position in three ranks with their 16-foot-long pikes pointing towards the enemy. Cavalrymen could not come close enough to use their swords on the pikemen, and when their enemy was judged to be sufficiently damaged the pikemen would shoulder their murderous weapons and charge. However in 1547 at Pinkie, just outside Edinburgh, when an English cavalry force encountered a resolute phalanx of pikes they did not try to drive a charge home but instead rode up, wheeled round and discharged their pistols into the faces of the kneeling pikemen. This manoeuvre so disorganized the pikemen that they failed to notice the approach of English halberdiers, whose shorter weapons were much more effective than pikes at close-quarter fighting. There were other factors contributing to the defeat of the Scots on that day, such as a bombardment from the sea, but the principal one was that they were fighting with out-of-date weapons, and their only asset was courage. Unfortunately for Scotland, that asset caused them to fight until their losses were enormous, when it would have been more sensible to withdraw, acquire new weapons and tactics, and return to fight another day.

As the search continued for greater simplicity combined with efficiency, the flintlock replaced the wheel-lock. The difference was that the trigger now released a spring which caused a hammer to hit a flint. In due course this technique would function *inside* the weapon, instead of at the side of it. There was some regret at the departure of the pike from the battlefield – though not on the part of the cavalry – but this was remedied by the invention of the bayonet. The first bayonets were of the plug variety – when a shot had been fired, the bayonets were plugged into the end of the musket. This clumsy device gave the Scots ample opportunity for revenging Pinkie when they encountered an English force at Killiecrankie in 1689. The force in fact included two Scottish regiments and one from Ireland, but the Highlanders were not concerned with details. The outcome of the battle was determined by the fact that, when the English force had fired a volley, it had no time to plug in its bayonets before a charge of wild Highlanders overwhelmed it. Their victory would have been greater still if the Highlanders had not stopped to plunder some baggage and thus enabled a few of the enemy to escape from the battlefield.

The defeated general, Hugh Mackay of Scourie, survived the battle and pondered for a long time over the circumstances of his defeat. Eventually he pressed for a weapon which could be used for firing or stabbing without alteration. It was not an entirely new idea, but it had not so far found much favour. The design was essentially simple; instead of being plugged into the muzzle, the bayonet was secured to the barrel by two rings; later it would be made easily removable by a clip-on mechanism. It was much too long at first, making it a cumbersome weapon, but over the years it became shorter and shorter. Bayonets have a psychological quality which makes them more effective than is justified by their destructive power: it is summed up by the words 'cold steel'. Men who wait unflinchingly through a hail of bullets and shells are distinctly unnerved at the approach of bayonets carried by apparent madmen. Bayonet drill is designed to inspire the attacker to heights of ferocity, while terrifying the intended victim into surrender (if he is lucky enough to be permitted it). Yet in fact the bayonet is a weapon of very limited firepower, and unless a bayonet charge comes in from very close quarters it can be stopped in its tracks by well-aimed shots. The modern bayonet is more akin to a dagger than the pike or spear which it replaced.

The next evolution in firepower was the practical breech-loading gun. The word 'practical' is important. Early experiments with breech loaders had been attempted in the fifteenth century, but the tendency was always for the charge to burst the barrel rather than project the missile. The true breech loader did not come into use until the mid-nineteenth century, and by that time great strides had been made in the design of projectiles. Among these improvements was the introduction of rifling – making cylindrical grooves along the barrel so that the bullet or shell would spin as it left the muzzle and thus travel with greater accuracy.

Breech-loading rifles changed the entire pattern of military deployment. In the days when muskets had to be reloaded through the muzzle between shots, the soldier who was reloading was so vulnerable as he engaged in his task that he had to retire temporarily from the firing line. He did this by taking paces sideways and backward, allowing the man immediately behind him to take his position in the front rank. This was practised as a drill manoeuvre, and in all armies what had begun as necessary movements soon became a ritual which was observed for its own sake. Undoubtedly, meticulous drill did improve discipline, bearing and co-ordination on the battlefield, but it acquired such a vested interest that it continued long after it had ceased to have any special relationship to firepower. Men were drilling to the eighteenth-century model as late as the First World War: even today ceremonial parades are based on drill movements which have had no practical application for over a century. Devotees of drill have made heroic efforts to preserve their embattled heritage. The substitution of the Self-loading Rifle (SLR), with its shorter length and awkward projection, for the Lee-Enfield .303 sounded the death knell to 'Slope arms', and the introduction of the even shorter rifles of today almost produced a crisis of confidence. However the drill sergeants have remained, as one would expect, steady on parade. Martinet, who was a drillmaster in France in the reign of Louis XIV, must be looking down – or perhaps, as some recruits who have cursed him heartily would aver, up – at the efforts of drill instructors to defy the passage of time.

A curious feature of most firearms is that no one seems to know exactly when they were invented or by whom. The percussion cap, which revolutionized firing, is thought to have been invented towards the end of the Napoleonic Wars by an American living in Philadelphia. Not perhaps surprisingly, in view of military conservatism, it was neither hailed with widespread enthusiasm nor adopted promptly. It is said – perhaps with truth – that the cold reception given to the percussion cap in the world at large was due to the fact that Europe was overstocked with flintlocks as a result of the Napoleonic Wars. That fact, combined with a natural reluctance to maintain large armies in the wake of a war which had lasted intermittently for twenty-three years, was a good enough reason. The armies fighting in the Crimean War of 1854–6 – British, French, Russian and Turkish – all used flintlocks, but less than ten years later, in the American Civil War, the flintlock was at long last declared obsolete. That same war saw the introduction of the combined cartridge. The American Spencer rifle fired a cartridge with the primer, the propellent charge and the bullet all within the same casing. They were known as rimfire cartridges, as they contained a ring of fulminate of mercury in the copper base. When the base was hit by a hammer, the bullet was launched by the propellant.

The nineteenth-century German 'needle gun' (more correctly the Dreyse rifle) pierced through the propellent to the primer, thereby ensuring proper ignition, and was extremely successful. However, like all weapons it soon revealed disadvantages: one was a slight blow-back of gas, which was extremely unpleasant for the user and caused fouling, the other that with the materials available at the time the needle was too slender to be robust and bent easily. Blow-back was largely eradicated when it was realized that the propellent and the projectile could be separated. Thus there came into being a weapon with a bullet which could be projected from a case containing the percussion cap and propellent at the rear. In the present century the gases produced by the explosive charges have been harnessed to reload the rifle and make it fire automatically.

One of the most famous rifles of the nineteenth century was the Minié, which took its name from its French inventor, Captain Claude-Etienne Minié. Although not produced until 1849, its merits were so obvious that it was widely used in the Crimean War only five years later. It had a rifled barrel which turned the 7/10-inch diameter bullet 1 inch in every 72. As the rifle was a muzzle loader, the bullet had to be narrow enough to slide smoothly down the barrel, but in order to pick up spin from the rifling of the interior of the barrel it needed to be a close fit on the way out. This purpose was achieved by making a conical hole in the cylindrical lead bullet. The propellent charge drove into this conical hollow, forcing the bullet outwards, while at the same time expanding it sufficiently to fit the rifling grooves tightly. The conical bullet went spinning on its way for 1000 yards with much more accuracy than its ball-shaped predecessors had done. Accurate fire at this range spelled the doom of not merely the cavalry but also most of the contemporary artillery. The former would be cut down before a charge had proceeded very far, and the latter would be picked off as they tried to load and fire their guns, most of which had a shorter range than the Minié. Though the Minié was not 100 per cent accurate, it was a great improvement on any weapon the infantry had used so far. It made possible one of the most famous events in British military history – the Thin Red Line.

The scene was Balaclava on 25 October 1854. There were three notable engagements on that day, and the last – but the least important from the military point of view – was the charge of the Light Brigade. The most productive were the charge of the Heavy Brigade, in which 800 British soldiers put to flight 3000 Russians, and that of the 93rd Highlanders who, a mere 550 strong, were all that stood between the Russian cavalry and the vital port of Balaclava. As the Russians bore down on the 93rd (who would later, with the 91st, form the Argyll and Sutherland Highlanders), those Scots who wished to stake everything in a death or glory attack on the Russians were steadied by the veteran Sir Colin Campbell with the encouraging words, 'Remember, there is no retreat. You must die where you stand.' Scots being Scots, his words were greeted with a cheer. The first volley from the Miniés of the 93rd was at long range: it checked the Russians but did not stop them. However, bullets arriving at that range came as a considerable surprise, and when a second volley followed soon afterwards the Russian commander decided that the firepower of the Thin Red Line (British troops wore scarlet: camouflaged uniform would not be adopted until later in the century, as a result of bitter experience) was too great to be risked. The encounter was really a test of nerve, but the Russian commander's failed first for, instead of spurring on a crushing charge which the 93rd would have had little chance of stopping, he wheeled away to the left. Campbell is reported to have said, 'That man knows his business', and obviously expected a flanking attack: he therefore ordered one more volley, and the Russian commander decided to fight again another day. Balaclava had been saved by the Thin Red Line, Scottish courage, a cool commander and – by no means least – the long range of the Minié rifle.

The Minié did not remain long in its original form, and was converted to a breech loader by giving it a different action (the 'action' is the mechanism which fires the gun). Nevertheless, it was soon replaced by the Snider, developed by an American inventor, Jacob Snider. The .577 Snider, adopted and used extensively by the British Army in the late nineteenth century, was followed closely by the American Springfield, with a slightly smaller calibre (diameter of the bore expressed as a decimal fraction of an inch), by the Chassepot in France and by the Mauser in Germany.

In 1845 a new explosive was discovered when cellulose was treated with nitric or sulphuric acid, making nitro-cellulose, which was manufactured at Faversham in Kent until an explosion wrecked the mills in 1847. Twenty years later it was made more stable by the addition of glycerine to the explosive and became known as gun cotton, from the original method of manufacture, which involved dipping cotton into a mixture of nitric and sulphuric acid. In 1859 Albert Nobel (later, ironically, founder of the Nobel Peace Prize) added an earth called kieselguhr, thus producing dynamite. Blended with nitro-cellulose and acetone, and produced in cords, it was called cordite. Further developments produced lyddite, so named from being manufactured at Lydd in Kent, TNT (trinitrotoluene) and amatol, which consisted of TNT and ammonium nitrate. Nitro-cellulose was first used in rifle bullets by the French in 1885, and had the advantage of being smokeless and more powerful. Smaller-calibre (.30) rifles could now be made; their range was longer and the weapons and ammunition lighter.

Clearly the need in the late nineteenth century was for some weapon which would reduce the interval between firing each shot, much of which time was taken up by reloading. Early experiments on feeding bullets into the gun by means of a lever were neither satisfactory nor reliable. However, in the meantime considerable progress had been made in developing a box magazine from which rounds could be fed into the rifle once a suitable repeater action had been designed. The successful box magazine was the result of the experiments of von Mauser in Germany, von Mannlicher in Austria, and the development staff of the Remington Arms Company in the USA.

Eventually the repeater problem was solved by the bolt-action Mauser, which, originally made for the Belgian Army, was soon adopted worldwide. The principle of the bolt action was that, when the bolt was pulled back, a round (bullet) was pushed up by means of a spring from the box magazine below. As the bolt was pushed forward and snapped downwards to lock it in place, the round was pushed up into the breech. When the trigger was pulled, the firing pin hit the percussion cap of the round, thus propelling the bullet. When the breech was opened again, a lug on the bolt hooked back the empty casing, which fell sideways to the ground, and the rifle was ready for the next round which had now been pushed up from the magazine. This process required a well cared-for weapon – dirt or fouling could and would cause it to jam. Not surprisingly, soldiers were taught that their rifles were their best friends, a thought which might not have occurred to them when they were cleaning them or drilling with them but which would become apparent when the enemy were approaching rapidly.

There had been many contributors to the development of the rifle, some with inventions, some with improvements. Two whose names seem to have disappeared into oblivion are Lee, a Scottish watchmaker, and Metford, an English inventor. The Lee-Metford, which they designed, was the precursor of the renowned Lee-Enfield which was produced at the Enfield factory from the 1890s onward. It was cut down to become the Short Lee-Enfield (SLE), and was a principal weapon in both world wars. In the Second World War it was supplemented and sometimes replaced by automatic or semi-automatic weapons, but for anyone with a good eye who wished to wound an enemy at a distance of a mile it had few equals.

The SLE .303 calibre is a legendary weapon. In 1914, when the Germans invaded France, they were met by a British force which the Kaiser had seen fit to describe as 'a contemptible little army'. 'The Old Contemptibles', as they came to be known, were small in numbers but were highly trained marksmen. They are said to have fired at fifteen rounds a minute with such steadiness and accuracy that the Germans assumed they were facing machine guns. Fifteen rounds a minute means three clips of five rounds in the magazine, and subsequent users of the SLE have wondered how they managed three clips without jamming; two clips is the average load, although of course three is perfectly possible.

The Lee-Enfield was greatly admired and coveted in countries where the British Army served. In order that British soldiers should not have them stolen from their sides while they lay sleeping in their tents, the rifles were often chained to their wrists or ankles. But this did not stop the expert thieves. On occasion they would arrive in the tent, their naked bodies greased and blackened so as to be neither visible nor holdable, and would position themselves, one at the soldier's throat with a knife, the other with a file to cut through the chain. If the soldier woke and realized what was happening he remained very still, perhaps encouraged to do so by the gentle pressure of the assassin's knife just above his jugular. In many countries much ingenuity was shown in manufacturing imitation Lee-Enfields, but the problem was always the bolt, which required a special quality of steel to stand up to the rigours of firing.

In modern war, when the opposing lines become static, as they did for almost the whole of the First World War, large numbers of men on each side are killed by what are generally known as snipers, but which the army prefers to call marksmen. Snipers are exceptionally good shots endowed with much patience. They are aware that, if they keep a close watch on the opposing parapet, sooner or later someone will incautiously raise himself, either to stretch from a crouching position or even to take a quick look towards the enemy lines. If he does so, he is dead: snipers rarely miss.

Snipers are hated by both sides, and often despised for their particular speciality; ironically, if a sniper kills a friend of yours you tend to detest all snipers, even your own. This is an unfair judgement. To be a successful sniper requires exceptional courage and endurance. In his lonely position, perhaps in a church tower or camouflaged in a tree, he is completely unprotected by his own side; and if he is discovered, his fate is likely to be unpleasant. When British parachutists landed on the night of D-Day they expected to lose men in frontal clashes, but they also lost men from snipers left in tall buildings.

In 1987 the British Army is due to receive the first consignment of a new sniper's rifle known as the Green Machine, said to be four times as accurate as the normal infantryman's rifle at long range. It has its own attached folding tripod which saves the user from having to find something on which to rest the weapon when he aims, and it also has an ingenious telescopic sight, produced by Schmidt and Bender of West Germany, with a built-in system for correcting the aim in cross-winds. One of the problems of hitting targets at long range is that under certain conditions the bullet will drift. The firer 'aims off' to correct his aim, but his job will be much simplified if this uncertain calculation is done for him.

Pistols and revolvers have never matched the accuracy of rifles, in spite of many attempts to make them do so. Samuel Colt took out the first patent for his famous revolver in 1835 in Connecticut, and this long-barrelled six-shooter soon became famous throughout the West. Other excellent pistol designs were produced later in other countries, but they never acquired the glamorous reputation of the Colt.

Revolvers are, of course, very useful weapons but their limitations are considerable. It is possible to fire a revolver accurately, but to do so requires unusual skill. Most people rely on the thought that if you can't hit a target with the first shot, one of the remaining five may be more successful. The revolver is at its best when its muzzle is pressed into the back of the victim's neck. Stories of cowboys advancing towards each other, flanked by expectant and admiring friends, and then firing from the hip with killing accuracy are mainly the work of creative fiction. This does not mean that such contests did not occur occasionally, but the average cowboy gunman was not addicted to sporting contests with even chances: his idea of a successful revolver shot was one which hit the unsuspecting victim in the back.

The Colt .45, being large, heavy and cumbersome, was replaced by the smaller .38. These were carried by officers in both world wars, not so much with a view to inflicting damage on the enemy but to ensure that nobody changed his mind about going over the top when the whistle was blown. Death in No Man's Land was possible, even probable, but death if you tried to stay behind was certain.

For emergencies, in which an instant shot might be necessary, a revolver could be cocked by pulling back the hammer so that it would fall instantly when the trigger was pressed. Having the revolver cocked like this made it highly dangerous, for a knock could cause the hammer to fall and the bullet, with its 'great stopping power', as the army would describe it, to find the nearest target. Slower but safer was the method which caused the barrel to revolve and the hammer to be drawn back before it fired.

The older type of manually operated pistols and revolvers (not all pistols were revolvers) were durable, heavy and not prone to jamming. When they did jam it often led to fatal accidents, for the user could easily forget in which direction the barrel was pointing as he tried to make the gun function.

Although revolvers have had widespread use in armies in the present century and have not been entirely discarded even today, automatic pistols were developed as early as the nineteenth century. The principle was simple enough: when the trigger was pressed the first bullet was fired, and the gas from that explosion was used to put the second bullet into position and fire it. The problem with this type of pistol is that, although it is possible to aim accurately with the first shot, the subsequent shots are extremely inaccurate – often high – as the pistol jerks wildly.

More success was enjoyed with pistols which made effective use of the recoil. From the beginnings of gunpowder one of the great problems had always been to absorb the recoil, which drove the entire weapon forcibly backwards as the round travelled up the barrel. Among the better-known weapons which used the recoil principle were the British Webley and the German Luger and Mauser. Of these the Luger became the most famous, probably because of the success of the long-barrelled model. This was loaded automatically, but fired by single shots. It was originally designed by an American named Borchardt, but was developed by the German after whom it is named. Another notable pioneer in this field was Louis Schmeisser, a German designer who, towards the end of the First World War, produced a reliable weapon which fired a 9-mm bullet.

These developments in pistols ran closely with the advances in practical machine guns and sub-machine guns. They were also concurrent with such designs as the American Smith and Wesson repeater rifles, which Horace Smith and Daniel Wesson had first produced in 1855. Surprisingly, the first machine gun that worked had been invented as early as 1718 by James Puckle: it was rotated by hand and fired by flintlock. However, it was too cumbersome to be used on the eighteenth-century battlefield, and not until a hundred and fifty years later was a fully automatic machine gun produced which met the military requirements of the time. This was the original Maxim gun, invented by the American-born Sir Hiram Stevens Maxim, which worked on the principle of harnessing the recoil. Maxim also invented a form of smokeless powder to go with it. His nephew, Hiram Percy Maxim, added a silencer which he had developed in the course of trying to muffle the noise made by petrol-driven vehicles. Unfortunately for him, his gun silencer brought him widespread opprobrium because of the fear that it would be fitted to the guns used by criminals; the fact that it could not be fitted on the type of weapon criminals were likely to use did not stop its being banned in many US states. Both Maxims were prolific inventors: Hiram Percy built electric cars and Sir Hiram invented mousetraps, automatic sprinklers and even a steam-powered aeroplane which succeeded in flying. Sir Hiram, who had become a British citizen in 1900, was knighted by Queen Victoria in the following year. His own manufacturing company was, very appropriately, merged into Vickers, producers of a wide range of automatic weapons. The water-cooled Vickers, based on the Maxim, was an exceptionally reliable weapon.

The Gatling was another American invention; it preceded the Maxim but worked on a different principle. Richard Jordan Gatling patented his crank-operated multi-barrel machine gun in 1862. He had qualified as a doctor, but preferred inventing agricultural machinery to practising medicine; as his farm implements made him wealthy and famous, he seems to have been wise to follow his inclinations. The Gatling fired a metal cartridge every time the handle was turned: loading, firing and ejecting followed in rapid succession. The special virtue of the Gatling was that it never overheated, because each shot used a fresh barrel until the complete rotation had been made. Its ability to fire a stream of .58 bullets made it an awesome weapon, but its popularity was less with the US Army than with the armies of other countries. Gatling continued to improve his hand-cranked guns and produced both gas-operated and electrically powered models. However, the fact that by 1900 the latter could fire 3000 rounds a minute did not shake military resistance to change and it was not adopted. In consequence, although Gatling himself died in 1903, his hand-cranked guns were being produced, and sold world-wide, as late as 1911.

Yet another hand-cranked weapon was the French Hotchkiss, developed in the late nineteenth century. The Hotchkiss had considerable stopping power but was so heavy that its best use was considered to be when mounted on ships, and it performed well on gunboats. However, the French Army used the Hotchkiss very successfully in static defence during the First World War. Although air-cooled, it contained an ingenious device by which after each burst of firing the breach remained open, greatly assisting the cooling process. In the Russo–Japanese War of 1904–5 the Russians used British-manufactured Maxims, the Japanese French Hotchkisses.

Although Britain had done much to perfect and popularize the machine gun, the armies which formed the British Expeditionary Force in the First World War were extremely badly and unimaginatively equipped with these essential weapons. This was due partly to poor planning and partly to inherent military conservatism. A newly arrived subaltern recalled one of his officers saying scornfully, ‘I’d like to see the machine-gun which could stop a cavalry charge,’ and lived long enough to see it happen. The British Army was reluctant to supply machine guns to rifle companies, which, they believed, were better off without having to lug around such unwieldy weapons; they preferred to employ them in a Machine Gun Corps, to be called upon when necessary. German policy was to spread machine guns evenly, which gave them enormous firepower when it was needed to check infantry attacks; proof of the success of this policy was demonstrated only too well at Loos, the Somme and Passchendaele, to mention but a few of the battlefields on which the Germans scythed down oncoming British infantry in swathes.

A major problem with all the early versions of the machine gun was its weight. The French *mitrailleuse*, also developed in the nineteenth century, had twenty-four barrels and altogether weighed over 2000 lb. The later French weapons were lighter, but still cumbersome. However, experience of trench warfare convinced all the belligerents that there was a vital need for a machine gun light enough to be carried by its firing team, or even by one man. The result was the Lewis gun, rapidly adopted and much cherished by the British; its creator was a US Army officer with the very appropriate name of Isaac Newton Lewis. He was a versatile inventor, being a pioneer of electric car lighting systems (as opposed to oil lamps) and of generating electricity by means of windmills, but his light machine gun, patented in 1911, had been rejected by the US Army. In Europe, however, he found a very different reception for his gun which, originally manufactured in Belgium, soon moved to England to be merged with the Birmingham Small Arms Company (BSA). BSA manufactured Lewis guns in thousands: they were popular not merely in the Army but equally so in the Royal Flying Corps. Finally, they were grudgingly accepted by the US Army. Contemporary with the Lewis gun were the French Chauchat (see Chapter 4) and the American Browning.

John Moses Browning, born in 1855 in Utah, was a precocious boy who made his first gun at the age of thirteen; when he died in 1926 his guns were world-famous. At the age of twenty-four he sold one of his rifles to the well-known Winchester Rifle Company. (Oliver Fisher Winchester had been a manufacturer of men's dress shirts before he created the Winchester repeater, 'the gun which won the West'.) Browning designed many guns and sold them to famous companies such as Winchester and Remington, but rarely bothered to claim the product as his own. However in 1918, when he sold his .30 gas-operated automatic to the US Army, it became well known as the Browning. Subsequently the design was modified to make the gun larger or smaller, but substantially the same weapon remained in use until long after the Second World War. The British SAS made good use of .50 Brownings in the Western Desert in 1942.

The SAS Regiment, which eventually did much to develop hard-hitting weapons, began its life in the Western Desert in 1941 when German forces were poised to sweep away the British Army defending the Egyptian frontier and then advance to capture Alexandria, Cairo and everything strategically important in the Middle East. Not the least of the German assets was the ability of their bombers to fly faster than any fighter the Royal Air Force could put up against them. David Stirling, a Commando whose regiment was in the process of being disbanded owing to the changed military situation in the area, decided that, even if it was not possible to destroy German aircraft in the air, well-armed raiders might perhaps succeed in doing so on the ground. With a few like-minded colleagues of adventurous and ingenious disposition he decided to venture over some thousand miles of desert, advance by night on the German airfields, and blow up their machine and petrol supplies. After an initial attempt at parachuting in had been turned into a disaster by hurricane-force winds, Stirling decided that truck and foot were the only answer. The trucks were 30-cwt, carefully adapted for desert travel, equipped with sun compasses, steel channels for getting out of soft sand, and water condensers: they could go for 1100 miles without refuelling. Initially their armament was not very powerful, but it was increased dramatically when Stirling found some Vickers K machine guns which had been meant for the now obsolete Gloster Gladiator aircraft. The old and slow Gladiators had been equipped with two Lewis guns and two Vickers. By the time Stirling acquired the Vickers guns he had also obtained jeeps, which were more suitable for his purpose than the 30-cwt trucks. The Vickers guns were mounted in pairs at the front and rear of the jeeps, and even without the addition of the .50 Browning the effect was devastating. Each Vickers could fire 600 rounds a minute, and the accompanying noise was almost a weapon in itself. The range of the Vickers was 3000 yards, much the same as the Browning.

One of the SAS's most devastating raids was on Sidi-Enich airfield, which was crammed with German planes of every variety. By the time this raid was planned in 1942 the Germans were well aware of the new threat to their ground installations and had made appropriate dispositions. Searchlights played over the perimeter boundaries, and armoured cars were on standby. The SAS decided to eliminate Sidi-Enich with overwhelming firepower. Using eighteen jeeps, each equipped with four Vickers Ks, the SAS suddenly drove into the middle of the airfield in a carefully rehearsed attack. Having cruised down the middle, destroying everything it encountered – including an aircraft which had unhappily for itself chosen that moment to come in and land – the SAS party paused, reloaded and then circled the perimeter in another devastating tour. Every plane on the airfield was destroyed and with them went huts full of stores and petrol. The SAS lost three jeeps, had one man killed and suffered badly from scorching. It was a classic example of what surprise, coupled with overwhelming firepower, can achieve.

This combination was, of course, the secret of the many successes of the SAS. Any weapon which contributed to this end was brought into service and, if necessary, modified to give it even more killing power. Much use was made of captured enemy weapons when these seemed particularly useful for SAS purposes. Luger 9-mms and Walther P.38s were popular weapons, both having an *accurate* range of 75 yards and a maximum range of 1200. The Schmeisser machine pistol was more powerful. In its early stages this model of the Schmeisser, developed by Louis Schmeisser's son Hugo, had been built outside Germany in order to evade the specifications of the Versailles Treaty, but by 1939 was in full production at home. It held thirty-two cartridges in its magazine and could fire at a rate of 500 per minute with an effective range of 150 yards. The greater the speed of fire, the greater the need for accuracy, for there is no chance to correct a false aim before the entire magazine has gone. The interval between the last shot and replacing the magazine can well be fatal to the user who, if he had had a weapon with a slower rate of fire, could perhaps have corrected his aim before exhausting his ammunition. Military history abounds with stories of men who fell easy victim to an enemy with an inferior weapon because they themselves had used up their own supply on the wrong target. Rapid rates of fire can also be devastating to one's own side. When soldiers are advancing under cover, it is surprisingly easy to mistake friend for enemy. On several occasions in the Malayan jungle in the 1950s British troops stumbled into ambushes which their own regiment had carefully laid to entrap the enemy: every mistake was fatal.

An advantage of an automatic weapon is that it not merely possesses a faster rate of fire but can be used without the firer making an observable movement. Flash from the muzzle may be a problem, but flash eliminators of varying quality are available. A notable disadvantage is that the weapon creates problems of logistics by the speed at which it uses up ammunition. The famous Thompson submachine gun, the 'tommy gun' (described in Chapter 5), is remembered by Second World War soldiers partly for its power at short range but more so for the weight of the ammunition which it blazed away with much less effort than was required to transport it.

As mentioned earlier, the user of an automatic weapon does not have to perform any other movement than changing the magazine. This gives it a huge advantage over its bolt-action predecessors, for every time these older rifles were loaded the user's arm had had to make a rapid movement. Movement is instantly observed: one reason why recruits are drilled to make them stand motionless on parade for long periods is that the ability to stand, or otherwise remain completely still, is an essential battlefield requirement: 'One move and you're dead', as people tend to say in gangster films. How visible the slightest move can be when all else is still may be tested by observing recruits as yet inexpert at drill. Having made an incorrect move, the unwise man – or woman – may gently try to correct it and adopt the same position as all the others. If he had stayed with his error he would not have been observed except by those closest to him, but the slightest movement immediately gives him away. Small wonder that the barked command 'Stand still!' or its foreign language equivalent has echoed over parade grounds worldwide.

4

The Heyday of the Automatic

Although various forms of automatic weapons were produced earlier, the principle of automatic fire has never been as widely applied as it has come to be in the late twentieth century. A weapon is now required to do everything for itself except think – and at times it appears to be doing that too. Automation is, wherever possible, combined with miniaturization, a process in which the silicon chip plays a vital role.

The fact that armies would turn towards smaller automatic weapons became obvious in the Second World War. Automatic weapons such as machine guns and repeater rifles, in existence already, were, however, considered to be more an addition to firepower than a basic article, and progress in the early stages was neither easy nor even. Britain used the tommy gun for close-quarter work: its problems have already been described. For longer ranges the Lewis had now been replaced by the Bren, a gun of Czech design originally produced by the Skoda arms factory at Brno.

These were the first light machine guns the world had seen, and they were soon manufactured under licence by other countries. The Bren was portable, as it weighed only 20 lb, but it had a range of up to 2000 yards, over which it fired 600 rounds a minute. Low-flying aircraft, intent on a little ground strafing, could find themselves in serious trouble if they encountered a Bren, but it took a steady nerve to aim a Bren accurately at a plane which was screaming down on the gunner and spraying bullets all around him. Nevertheless there were plenty of owners of steady nerves, and some even fired the Bren from the hip. The Bren was originally manufactured to take .303 rounds, but was later converted to 7.62-mm, which is slightly smaller but considered to have adequate stopping power. The gun was greatly valued, but required understanding and careful handling; it could cease to operate for a variety of reasons, all of which needed to be clearly appreciated by the user. To prevent overheating, barrels had to be changed after every ten magazines. The idiosyncrasies of weapons such as the Bren are remembered long after the user has ceased to operate one, and in many cases long after the weapon itself has gone out of service. There are civilians all over the world who, if required to do so, could run through the action required to clear stoppages in the weapon they last handled over forty years ago. What was once painstakingly learnt eventually becomes an instinctive response to be followed in darkness or light, in heat or cold, even when seriously wounded. It becomes indelibly etched on the memory, like the number given to the soldier when he joins and which is only changed if he is commissioned.

The German Army was equally well supplied with Brens, for the Nazis had take over the Skoda factories in 1938. They fired 7.92-calibre ammunition from a thirty-round magazine. The gun was introduced to the German Army after it had undergone various modifications; this weapon was twice as heavy as the British Bren, and when used as a machine gun was belt-fed. A British version of this heavier gun, known as the Besa, was used as a tank machine gun. An even larger version of the Czech Bren was the ZBvz/60, a powerful brute with a 15-mm calibre, a weight of 121 lb and a rate of fire of 420 rounds per minute (rpm) from forty-round belts.

Germany had a powerful armoury of machine guns in both world wars. The principal killer weapon in the First World War had been the Maschinengewehr 08, which was built at Spandau, and for that reason is usually referred to (by the British at least) as the Spandau. This too had a calibre of 7.92 and fired up to 450 rpm from a belt holding 250 rounds. Its weight – 137 lb – was a drawback. Unless it suffered a direct hit, the 08 Spandau was virtually indestructible. Although most 08s were confiscated or destroyed in the 1920s to conform with the provisions of the Treaty of Versailles, some were permitted to be retained by the small German Army and police force. They were still serviceable in 1939, and many were still being used in 1945.

As mentioned in connection with the SAS, guns designed and

manufactured in one country very often finished up being used by that country's enemies; sometimes they were recaptured by their original owners. Colonel Norman Berry of the Eighth Army was surprised to find that a very effective German anti-tank gun which they had captured was in fact a British-manufactured 3-inch anti-aircraft gun which had been fitted to a Soviet gun carriage. Sent to the USSR on one of the Arctic convoys, it had been captured from the Soviets in one of the earlier battles; then, having been serviced, it was sent to the Middle East. All this had happened within four months.

In 1919 Germany had handed over 08s to Poland as part of their reparations bill. When the Soviets took over half of Poland in 1939 they added the 08s to their own stocks. Many of them were recaptured by the German invasion forces which entered the Soviet Union in June 1941. The Soviets also used a number of German machine guns which had been sold by Germany to Portugal between the wars and were subsequently sold again to make a profit. A less salubrious story concerns the MG 35, which was a Swedish-designed gun. Production was started in Germany after 1935 (when Germany was openly rearming, having defied the Versailles Treaty), but the German Army did not wish to adopt it as it stood and the only customers were the Waffen SS. The Waffen SS, however, did not consider it good enough for their own use but found it useful to supply to the regiments they had recruited in countries they had occupied, such as Norway, Belgium, the Netherlands and France.

Another example of dubious neutrality was provided by Switzerland, whose Solothurn Company was owned by the large German armaments manufacturers Rheinmetall-Borsig. The weapon concerned was the Solothurn MG 29, which had an advanced design feature for its year, 1929, in that it could be fired either as a single shot or as an automatic, according to how the user pressed the trigger. The German Army was not sufficiently impressed with the gun's qualities to wish to adopt it, but the Austrian and Hungarian Armies took a more favourable view. However, after a few modifications by Rheinmetall-Borsig the weapon found Wehrmacht favour and proved very successful as the MG 34. It fired a 7.92 round at 800 rpm. The Swiss had no comment to make on the fact that, although a neutral country, they had allowed a German armaments giant to control one of their factories in defiance of the Treaty of Versailles and develop a machine gun of Swiss design. Doubtless it was all done for neutral and peaceable purposes. The Germans may have had reason to regret their involvement with this gun, for many found their way to the resistance forces and were used against them.

However, good though all these guns were, the Mauser MG 42 was considered to be better than them all. It could fire 7.92 cartridges at the devastating speed of 1500 rpm. Ammunition was held in belts of 130 rounds. Not the least of its virtues from the user's point of view was its lightness: it weighed only 25.5 lb. An improved version of this gun, the MG 45, was just being introduced to the Germans when the war ended.

The popularity of certain types of gun, which causes them to be preferred on some occasions to weapons with a better all-round performance, is often the result of what is called 'the human factor'. A popular Americanism nowadays is the term 'user-friendly', meaning easy and satisfactory to use. Some guns were distinctly not user-friendly. Sometimes their weight made them unpopular, but more often it was their habit of jamming at inconvenient moments, or even apparently attacking the user with sharp projections. Designers of military equipment do not, unfortunately, have to carry their creations through trenches, up mountains or on slippery jungle paths, and therefore remain happily unaware of the damage that a sharp corner or awkward shape can do to the carrier's body or morale. A gun which seems to have owed its popularity to being user-friendly is the Danish Madsen, originally a .303-calibre gun with a capability of 450 rpm, but which weighed only 20 lb. It was used by many countries including Britain for home defence units, mainly because it was adequate and reliable and did not suffer from stoppages which might baffle the unskilled user. In a very different category was the French Chauchat, which first came into use in the First World War. Its only virtue seems to be that it was light (20 lb); it fired an awkward calibre bullet (8-mm) with a poor output of under 300 rpm. A more successful venture was the Chatellerault range. In defence of the French guns it needs to be said that after the Germans had occupied France in 1940 they were hardly likely to encourage a French armaments industry working under its own power. Instead, they confiscated guns which they thought would be reasonably satisfactory as back-up weapons against emergencies such as German arms factories being put out of action.

The Italians, who were allied to Germany in what they called the

Rome–Berlin Axis, were not given as much encouragement in their arms manufacture as they might have hoped for. As a result they had rather less success with their products than might have been expected. Large numbers of *mitragliatrici* were made by Fiat for both world wars, but in neither did they do particularly well. Perhaps the most successful weapon was the Mitragliatore Breda, which had a 6.5-mm calibre and a rate of fire of up to 500 rpm. This stood the test of the campaign in the Western Desert reasonably well and caused no slight inconvenience to the British Army. (The desert, as might be expected, bore very hard on weapons, particularly automatic ones: the universal abrasive dust penetrated to all parts and there were inevitable problems from overheating.) The Germans were not above using Italians guns in certain circumstances: the Breda Modello 31 was a moderately successful tank machine gun, and the 37 the best of the heavy infantry machine guns.

As mentioned earlier, the Japanese used French Hotchkiss machine guns when they defeated Russia in 1905, and when they produced their own guns in time for the First World War it was no surprise to find that they bore a close resemblance to their French predecessors. The Taisho II, with a calibre of 6.5 mm and a rate of fire of up to 500 rpm, acquired a reputation for being a more efficient weapon than it was because the Japanese transported it to unlikely and often exposed positions and were perfectly prepared to throw away their lives in order to keep it operating. The kamikaze principle, which became well known in the Second World War from the suicidal crash landing of aircraft on to targets, was not limited to pilots. The quickest and most honourable way to Nirvana was to die in battle, which made the Japanese very difficult opponents. As their religious fatalism was accompanied by a marked desire to take most of their opponents with them on the final journey, they were formidable indeed. A weapon handled by a Japanese was likely to do more than its fair share of killing, unless eliminated by superior firepower from the other side early in the proceedings. Japanese soldiers would stay for days in ambush positions, though completely cut off from supplies; being wounded made no difference. Enemy tanks were stopped by Japanese soldiers diving under them with a pack of explosives strapped on their backs. Before the Second World War Japanese military ability had been seriously under-rated.

The Taisho II machine gun was also called the Nambu, after its designer, and many of them are still in use in South-east Asia today. The reason for the wide distribution of this machine gun was not a successful arms sales policy – when the Emperor ordered his armies to lay down their weapons, soldiers still serving overseas handed them to the guerrilla groups dedicated to establishing Communist governments in Burma, Malaya, Indonesia, China, Korea and Vietnam. By far the largest number eventually found their way into Vietnam. In view of the skill shown by post-war Japanese industry, it seems surprising that Japanese machine guns were not better made than the products of other countries, although many of them were an amalgam of foreign designs. However, the Japanese did produce a machine gun with one unique feature: the 6.5 96 could be fitted with a bayonet. This fact may seem almost bizarre to Western eyes, but it illustrates perfectly the dedication with which the Japanese set about their military task. When the machine gun had fired every available round it could still be an offensive weapon, difficult though it must have been to use with its weight of 20 lb.

Identifying the origins of a particular gun is not usually easy and is often surprising. Thus the Indian Army Vickers Berthier (VB) was a British gun manufactured to a French design in the 1920s. It was modified at various stages, but its outstanding reliability combined with powerful strike capacity, .303 calibre and 600 rpm made it a favourite gun with all its users. Eventually the rpm was pushed up to 1000. VBs were used on aircraft, and for defence against aircraft they were used by mobile raiding columns, notably the SAS. They were used in field armies and in home defence forces. The VB was one of the most versatile machine guns ever produced, yet it had begun as the Berthier, a product of a not very successful French armaments industry.

Considerable confusion over identifying guns occurs when captured models are renamed by their new owners. This, of course, can apply when guns have been bought from a foreign manufacturer. The Germans pressed into service any Vickers guns they could lay their hands on, renaming them the MG 230. An interesting gun to identify was the 7.92 Besa tank machine gun, a useful weapon which could fire up to 750 rpm. This gun, as seen earlier, was a development of the basic Czech ZB design which had been incorporated in the British Bren. The Besa was widely used in British armoured vehicles, and whenever these were captured the Germans removed the gun and used it themselves in their own tanks.

However, no gun appears to have been adopted by so many nations as the ubiquitous Lewis. Astonishing though it may seem, this First World War design was used by every belligerent army in the Second. It is credited with one-fifth of all the kills made by British pilots in the Battle of Britain. The presence of a Lewis gun on an airfield, ship, camp or harbour was a very considerable deterrent to any unwelcome intruder – whatever his nationality.

The Soviet machine gun, the Pulemet, made no attempt to disguise its debt to the Maxim design: thus when these weapons were captured in the Second World War the Germans had no difficulty in using them. The Soviets seemed less concerned with weight than were other nationalities, but later they produced an excellent lighter design of their own, the Degtareva Pekhotnii (DP), which fired a 7.92 round at 550 rpm and weighed only 26 lb. This gun has never gone out of service; its essential quality is the simplicity of its design: there is little to go wrong. Precisely the same principle was applied in their AK-47 assault rifle, used worldwide, of which more will be said shortly.

Although the machine gun is far from being the most destructive of modern weapons, the fact that it is a very personal one makes its firepower particularly awesome. Bombs, rockets or heavy artillery can kill large numbers, but their users are far removed from their victims. Machine gunners, on the other hand, select their targets and know very well what success they are achieving. They see their targets, living men, scythed down as the machine guns methodically traverse their lines. The Germans who desperately tried to break through the defences of Stalingrad, Leningrad and Moscow were cut down by equally determined, and perhaps even more desperate, Russians. The sound of the Soviet DPs was as familiar to the Germans as the MG 42s were to the Russians. Sometimes a machine gun pumps away methodically and remorselessly as if it has all the time in the world and is going to use it; at others it seems to explode in a frenzied burst in which individual shots are barely distinguishable, if at all.

Although the Soviets were quick to copy the successful designs of other countries, they showed that they were fully capable of taking ideas from various sources and blending them into an outstanding weapon. Such was the Goryunov 1943 G, a gas-operated, air-cooled gun which gave reliable service through the latter years of the Second World War and for many years afterwards.

Last, but not least, comes the American contribution to this gallery of weapons of instant death. One of the first of their guns to win widespread approval was the .30 Marlin, a gun which began its service life with the US Army in 1917 but was still being used as an anti-aircraft gun in the Second World War. Brownings, often known as BARs (Browning Automatic Rifles), with a 7.62 mm calibre and a rpm of up to 500, were the basic weapons of the US forces. The term 'rifle' seems rather unsuitable for a weapon weighing over 19 lb.

The Browning .30 M1917, the number being that of its first year of service, probably holds the record for the longest period of use experienced by any machine gun. The gun which was supplied to the US Army for the closing stages of the First World War had in fact been designed twenty years earlier, in the nineteenth century. Whether its long and far-ranging life was due to its innate superiority over any comparable machine gun or merely to the abilities of American arms salesmen is an unresolved question. With some modifications it was used as the main machine gun of the US Army during the Second World War – and by other nations too, when they captured them.

Other modifications of the Browning produced the M1919, an anti-aircraft gun which could also be mounted on armoured vehicles. From this it became the M1921 and, later, the notorious .50 Browning, mentioned earlier. In this heavy role (.50 inch is 18.7 mm) it had nearly twice the killing power of the lighter models.

Later modifications have produced even more powerful models of this long-serving basic design. If guns based on the original are still being used in ten years' time – and it seems likely that they will be in one or other of the world's many conflicts – that design will have been in use for over a century. The M60, the machine gun which has replaced the .30 Browning, is expected to remain in service for at least a decade.

In machine guns, as with other manufactured goods, there comes a stage when the optimum point of development has been reached. The first machine guns were unduly heavy by modern standards: they were efficient, but their weight, rate of fire, accuracy and reliability all needed development or modification. Their ammunition was by no means entirely satisfactory. Over the years the discovery of new alloys, explosives and methods of manufacture has reduced weight, increased killing power and improved reliability. For a long time the fearsome capabilities of machine guns have been all too well known, and no doubt they will be in service somewhere in the world for many years yet. But in terms of firepower they belong to a past era, a period when armies fought by attacking, or holding, entrenched positions on battlefields. It is difficult to visualize that sort of tactical approach on the battlefields of the future. The machine gun accounted for most of the losses in the First World War and it also took a heavy toll in the Second. It will no doubt continue to do so in minor wars. But if ever the world should be so unfortunate as to see another major conflict, it will probably be decided by a different range of weapons.

Signs of future change were to be seen early in the Second World War. In the machine age it seemed ridiculous that any form of repeater rifle should still be manually operated. Futhermore, the rifles of the First World War, many of which were still in use, were heavy, cumbersome and large enough to be conspicuous. The Germans made early experiments but found that the lighter automatic weapons lacked the range and killing power of the manually operated rifles. The answer was eventually found in the 7.92 Kurz round, fired by the MP-44. It attained a range of 1000 feet which, though less than that of the standard rifle, was adequate for most purposes. These new model automatic weapons became known as assault rifles. The MP-44 proved its worth during the Second World War, and after modification the design was adopted by the Soviet Army in 1951. It became the AK-47, which is now used worldwide. Like the MP-44 it is gas-operated and can fire either fully automatically or semi-automatically, according to the demands of the situation. It is light to handle, weighing under 11 lb, but the main virtue of the AK-47 is that it is a simple, relatively trouble-free weapon. It was soon adopted by the Chinese Army and forces from the Eastern bloc, and was subsequently distributed in huge quantities to guerrilla groups whose political ideas were either already Soviet orientated or seemed likely to become so. By 1987 the general public was only too familiar with pictures of guerrillas, urban terrorists and even children brandishing their AK-47s.

There is no doubt that the AK-47 is a very good rifle. It fires a 7.62 cartridge, which is said to be based on the German Kurz 7.92; the detachable, curved box magazine holds thirty rounds. AK-47s were widely used by the Vietnamese forces during the Vietnam War, and stood up to rigorous conditions remarkably well.

However, the Soviets had no intention of letting their arms development stand still, even though the AK-47 seemed to meet all requirements very satisfactorily. There was one notable disadvantage to the new rifle, but it was quickly eliminated. The first models had wooden butts and, to make matters worse, the wood was of surprisingly poor quality. In eradicating the fault the designers managed to make a considerable improvement to the rifle. The wooden butt was replaced by a metal one which would fold underneath, making the weapon easier to conceal without affecting its use. The AK-47 was designed to be fitted with a bayonet for use in certain situations, such as running out of ammunition, or when intimidation rather than destruction was required.

However in the 1960s the American Armalite AR15 (also somewhat confusingly called the M-16) proved very successful with a smaller-calibre bullet than had been in use previously. The calibre of 5.56 was found to be more than equal to the 7.62 round fired by its forerunner, the M-14. As the M-16 weighed 3 lb less than the M-14, as well as firing a smaller bullet, the US Army had no hesitation in adopting the new weapon. But even the M-16 had problems, as will be seen later.

The move towards a smaller-calibre rifle had not passed unnoticed in the USSR, although no hint of their thinking was allowed to leak to the Western world. Suddenly Soviet awareness of current developments was disclosed in the mid-1970s when they produced their own version – the AK-74. This, even with a loaded magazine, weighs less than the M-16. It fires a 5.45-mm round and has an effective range of about 550 yards. The latter, though some 300 yards less than that of its American rival, is adequate for most purposes. It is extremely accurate, but the degree of muzzle flash looks to be a notable disadvantage. For the basic weapon, as opposed to the model issued to parachutists, the manufacturers have reverted to a wooden butt, presumably of more durable quality than that fitted to its predecessor.

The development of modern rifles shows very clearly the dilemmas of weapon designers in the twentieth century, particularly in recent years. If the enemy has a similar type of weapon, ideally one's own should have longer range, greater accuracy, a lighter weight, more hitting power and less appetite for heavy ammunition. It should also be capable of enduring extremes of heat and cold, dust and mud, and any neglect or misuse that its owner inflicts on it. In practice, all designs are a compromise. The range and hitting power of a weapon may be increased, but it will probably involve a corresponding increase in the weight of both weapon and ammunition. Even if it's still not too heavy, it may be bulky and awkward to handle. If it is too light, its stability may be affected; if the rate of fire is extremely rapid, it may exhaust available supplies of ammunition too quickly – particularly in the hands of inexperienced, nervous or braggart troops. The designers may produce an excellent weapon but find it is unpopular because its potential users need too much training to master its complexities. Finally, no one knows what a weapon's capabilities are until it has been tried under campaign conditions. It may perhaps be *too* accurate. In the 1950s' jungle campaign in Malaya a favourite weapon was a shotgun: when firing at opponents masked by jungle, a soldier was more likely to make a hit with a shot which scattered than with one which went accurately on its way to a fleeing target. Similarly, in the Second World war part of the charm of the obsolescent Vickers machine gun was the fact that it was not too accurate. There is little virtue in pumping out a stream of bullets if they all go precisely into the same small area: a wide arc of fire will claim more victims. The Vickers was an example of a weapon that appeared to be superseded, only for it to be found that the weapon's very defects had become virtues.

Observation of modern battlefields, such as in the Iraq–Iran War and in Israel's encounters with her neighbours, makes it clear that mobility and speed are more important than ever before. Target identification no longer depends on the unaided human eye, and calculations which formerly took minutes are now made within seconds; everything has advanced impressively except for one single feature: the human being who has to start and finish it all. He or she is more vulnerable than ever; even though layers of nylon armour may keep out a bullet, they will not keep out a shell, and if the soldier is not killed by bacteriological or chemical weapons he may still die in the old-fashioned way of simply being blown to pieces. The expenditure involved in all the vast array of items of firepower designed to destroy the frail human body is daunting, but even more impressive is the waste. In the First World War fifty thousand rounds of ammunition were fired for *every* man killed. This, it will be remembered, was the period when the only automatic weapons were machine guns, and there was a limit to the number used. With modern automatic weapons, which can fire 500 rounds a minute instead of five or ten, the figures for wasted ammunition are stupendous. In Vietnam, two hundred thousand rounds were fired for every enemy soldier hit (Pentagon figures). Nevertheless the search for faster-firing weapons must go on, though ultimately design in every weapon must take into account the degree of tolerance that the human body can attain under battlefield conditions. The soldier may have tanks designed for him which are faster and more efficient than any previous armoured fighting vehicle, but if riding in them makes him 'seasick' and unfit to fight the designer will need to think again. Some concessions have been made to the fact that the human body is easily damaged: soldiers are now equipped with 'ear defenders' to prevent their hearing from being irreparably damaged. There is also a wide range of protective clothing (known to the British soldier as Noddy suits) to protect troops against various forms of contamination. However, the general assumption must be that the human frame is at least as durable as the hardware the soldier is handling. Unfortunately there is plenty of historical precedent for the idea that the soldier is at times more expendable than the weapon he is using.

Although Britain had an excellent opportunity to develop a self-loading rifle during the Second World War, the chance was missed. This was not the fault of military conservatism or hidebound stupidity but was related to the facts of life as they were seen at the time. Experiments had been made with automatic rifles in the 1920s and 1930s, and they had not proved particularly encouraging. At the beginning of the Second World War Britain, which had just introduced conscription, had an enormously difficult task in trying to instruct hundreds of thousands of troops in the basic skills of weapon handling. Few of them had fired guns before, and fewer still were experienced with rifles. There were already large stocks of the Lee-Enfield .303, and all members of the regular British Army were familiar with it. It was, in any case, a very simple weapon to understand and to maintain. Familiarity with the weapon was instilled by drill in which the weapon formed an integral part. Virtually any member of the regular army could become a moderately efficient drill instructor. Infantry tactics were based on the fact that soldiers possessed .303s and would be able to use them efficiently. The weapon had many uses apart from killing the enemy: it could be used to make an emergency bridge, and it could even be used to make a splint. To have tried to introduce an unfamiliar weapon, even one with greater firepower, would have produced a chaotic situation. The possibilities – and limitations – of a new range of weapons were not known.

However, one opportunity *was* missed. In Belgium, which has always been well to the fore in arms design and manufacture, a designer by the name of Saive had made considerable progress with a new, efficient self-loading rifle. When in 1940 the German Army arrived in his country, Monsieur Saive promptly departed for Britain, taking his plans with him. Although he did much useful work in Britain during the next three years, he was not called upon to develop a rifle for general issue to the British Army. A few models were made at the end of the war, but did not impress the War Office sufficiently for the experiments to be continued. In consequence, Monsieur Saive returned to Belgium where he received more encouragement. His automatic rifle was improved, manufactured and sold to the armies of a number of countries in addition to his own. Its teething troubles were not over; the early versions tended to fire high after the first shot, but eventually his FN rifle, named after the famous Fabrique Nationale d'Armes de Guerre, was adopted by NATO. However, whether the FN self-loading rifle could have been perfected, manufactured and standardized to fire available ammunition in wartime Britain seems, at the least, highly debatable. The Belgian rifle was not adopted by NATO without fierce opposition. The armaments firms of many other countries, particularly Britain and the USA, would have liked to see their own patents adopted. But in the long run the Belgian model seems to have been a very sound choice.

Introducing a new weapon to an army gives rise to many problems. Apart from those mentioned above there is a whole range of tactics to be replanned. This may be excellent news in terms of efficiency, but it inevitably encounters reluctance among those whose entire knowledge and experience of weapons are based on a different concept. Furthermore, until everyone is equipped with the new weapon two types of ammunition will be in circulation. Ammunition, as experienced soldiers know, has a mind of its own. If a box of .303 ammunition can be delivered to a 7.62 gun it will do its best to achieve this end. In peacetime this is annoying: in wartime it is disastrous. During the Second World War saboteurs managed to find their way into railway goods yards and sidings and change labels from one box to another. Subsequently German gun crews in the depths of the Soviet Union were not pleased to find that instead of the box of .88

ammunition they desperately needed they had been sent a set of tractor parts. The problem would normally be put down to the incompetence of the supply and transport authorities, and it was bad for morale. Wrongly directed ammunition was sent back, adding to the congestion on roads already overcrowded with upcoming and downgoing trucks containing everything from rations to casualties. Normally plenty of supplies go astray without sabotage, and continued to do so.

However, although Britain did not develop an automatic rifle before 1945 various modifications were carried out on the SMLE to make it more efficient. The No. 4 of 1941 had better sights and rifling; it was replaced by the No. 5 in 1944 and the latter, with its shorter barrel and lighter weight, was not phased out of British Army service until the late 1950s. It was a particularly suitable weapon for the warfare in which Britain was engaged in Malaya at that time.

In spite of the satisfactory record of the No. 5 during and after the war, Britain was by no means unaware of the need to develop an automatic rifle for the battlefields of the future. A design known as the EM2, which was a product of the famous Enfield arms factory, seemed a very promising weapon apart from needing one or two small modifications, but the fact that it took a .28 round damned it for NATO adoption: this calibre did not match any ammunition in current use with the US Army. However, the EM2 was not an entirely wasted production; many of its features, including its calibre, would reappear in the SA-80 weapon which eventually replaced the FN SLR. The SA-80 is shorter than the SLR by 4 ins, but weighs 1 lb more. Its reduced length makes it an easier weapon for troops who are required to climb in and out of armoured personnel carriers (APCs), and the fact that the calibre is smaller means that the soldier can carry twice the amount of ammunition he could manage for the SLR. The standard NATO round is now 5.56-mm. The size of the new weapon caused some headshaking and heartsearching from the drill martinets of the day, but they bowed to the inevitable and came to terms with it.

Needless to say, although standardization of weapons and ammunition is essential, national aspirations to develop superior weapons remain unchanged. During the 1980s the French have produced the FA MAS, which has the standard 5.56 calibre but can also be easily adapted to fire rounds of different calibre. The Israeli Army has a 5.56 Galil assault rifle. The Galil has a thirty-five-round magazine, but, like the other assault rifles, can also fire grenades.

The German Army was scarcely likely to be left behind in weapons development. The FG-42 has already been mentioned. The Maschinenpistole (MP-44) was the ultimate modification by Schmeisser of the MP-42 and MP-43. It fired a large round, the 7.92, at 500 rpm, and had a range of 875 yards. The Heckler and Koch (HK33) had a more complicated history. This too was a Sturmgewehr (assault rifle); it fired a 5.56-mm round with impressive accuracy at 600 rpm. It weighed only 7.7 lb, could hold up to forty rounds in a magazine, and was only 37 ins in length. A possible disadvantage was its shorter range: at 400 metres it was only half that of the MP-44. Offsetting this shorter range is the fact that an assault rifle is not required to hit its target at long distance. The concept of warfare to which it belongs is that of mobile units being carried in armoured vehicles or helicopters to contested areas, and then overwhelming the enemy with concentrated fire. Static warfare is no longer feasible, for any strongly defended point above ground can quickly be eliminated by massive firepower. But in the end wars will be won, as they have always been won, by the infantryman standing on the last piece of disputed ground. Whether that scorched piece of territory will, in the future, be of value to anyone is a philosophical matter which has no relevance to this examination of firepower.

The assault rifle is of great value in another development of modern times – urban fighting. During the Second World War it was gradually appreciated that the best way to conquer a country is not necessarily to reduce its cities to rubble. The experience of pounding Aachen with bombs and shells convinced all students of warfare that destroying a town by explosion can make it more of an obstacle than it was when it was virtually intact. That was the lesson of Cassino too, where the destroyed monastery became a greater obstacle than the original building could ever have been, and also of Stalingrad and many other places. An undestroyed town will undoubtedly be a nest for snipers, machine gun posts and booby traps, but it is better to capture it and clear it with troops using assault rifles and similar-scale weapons than to reduce it to rubble, which will hinder the advance of armoured columns.

The United States have, as one would expect, been pioneers in the use of automatic weapons; the US Army was the first in the world to adopt a self-loading rifle as standard equipment. This was the .30 M1, otherwise known as the Garand, which came into service in 1930. It operated by gas and piston, but had considerable limitations in that the magazine capacity was only eight rounds; even these had to be inserted in two blocks of four each. Nevertheless the Garand proved a very serviceable weapon; it stayed in use until after 1950, by which time over five million had been supplied to the US Army by such diverse manufacturers as Springfield, Winchester and the Italian Beretta Company. The Garand had a fitting for a bayonet, had a range of 1200 yards and weighed 9.5 lb. However, long before this tough and enduring weapon was being phased out experiments were being made for a successor. This was the M.14, which fired a 7.62-mm round. The M.14 had a twenty-round magazine and could fire either single shot or fully automatic, though if it continued with the latter for too long the barrel tended to overheat and there was no provision for changing barrels. Nevertheless, it performed notable service in the war in Vietnam. It became obsolete with the introduction of the 5.56 round.

There has been inevitable confusion between the M1 rifle (Garand) and the M1 carbine. A carbine is a lighter, shorter weapon, originally developed for use by cavalry. Before the development of light, easily produced machine guns there was no practical alternative to the efficient but heavy tommy gun. American military forecasters in the 1930s could see the need for a light, short, easily handled weapon to arm drivers of vehicles and anyone else for whose job the Garand was too long and unwieldy, but the M1 carbine was not put into production until 1941. It has been said that the M1 carbine was more an extension of an automatic pistol than a smaller version of an M1 rifle, and the fact that it fired a short, stubby round supports that view. The M1 carbine did not stay unmodified for long. Its successor, the M1 A1, was the same in essentials, but a folding metal stock replaced the original wooden one. The M2 and M3 could be adapted for night fighting or even automatic fire, but even without these extras this .30-calibre weapon was so popular with the soldiers who were required to use it that seven million of them were eventually manufactured.

In the 1950s, when there was a general move towards fully automatic weapons, the United States began experiments with what was destined to become one of the most famous of all American weapons, the Armalite M16. At first it was simply known as the AR, and went through various numberings (AR10–15) as the model developed. By using aluminium and plastic in its construction, weight was kept to the minimum. When the M16 was developed it had the capacity to fire at 800 rpm over a range of 500 yards, nearly twice as far as its predecessor. In relation to its firepower it is astonishingly light to handle and it was a very popular weapon with US troops in Vietnam. Unfortunately, its general characteristics make it very attractive to terrorists and subversive groups, with which it has acquired considerable notoriety. Its principal fault is a tendency to foul easily.

An even more compact version of the Armalite is the weapon known as the Colt Commando (Colt are the manufacturers of the Armalite). It is shorter than the M16, with an overall length of 28 in as opposed to 39; its barrel is 10 in compared to the M16's 20, and although it has a slightly lower firing rate (750 rpm) it has the same range. The shorter barrel necessitates a flash hider, but if the soldier is working in a very confined space this addition may be removed, reducing the length even further. The Colt Commando may be said to bridge the gap between rifle and sub-machine gun, and it can be classed in either capacity.

Rifles of all types, varying from the primitive to the ultra-sophisticated, are in use all over the world. There are plenty of home-made weapons in the hands of guerrilla or terrorist organizations. Some are so crude in appearance that they look to be a greater danger to the firer than to the target. Even so, it is worthwhile remembering that a bullet from a weapon made of piping and stolen parts can kill a man just as dead as a rocket with a nuclear warhead.

5

Close-quarter Killers: Sub-machine Guns, Pistols and Grenades

Although the sub-machine gun (SMG) is generally thought of as a weapon of very recent vintage, its history goes back to the First World War. Hugo Schmeisser and his machine pistols have already been mentioned in Chapter 3, but the Italians were also early in this field with their Villar-Perosa. Unfortunately for the Italian armaments industry, the first sub-machine gun to be used on the battlefield was Schmeisser's weapon, the MP Bergmann, which had a 9-mm calibre, a reasonably light weight at 9.2 lb and a rate of fire of 400 rpm. Once the Bergmann trigger was pressed, fire was continuous until the magazine was exhausted. There is, of course, a tendency for even experienced soldiers to react to an alarming situation by a continuous burst of fire. This creates impressive noise and uses up an unnecessarily large amount of ammunition, but rarely has a decisive effect: in fact it may even be counter-productive. If the gun has an unusually large cartridge it will have 'great stopping power', but will place an almost intolerable load on the supply lines.

The inventor who designed and gave his name to the tommy gun was Colonel J. T. Thompson. To credit him with inventing it is probably to overstate his claim: he was Director of Arsenals for the US Army Ordnance Department, and as such he led a team of experts whose brief was to design and develop a practical sub-machine gun. Its original purpose was for close-quarter work in trench warfare, but the First World War finished before its capabilities in that role could be fully tested. An improved model was produced in 1921, but the American nation was not particularly militarily minded in the aftermath of the war and orders were small. However, both Army and

Navy took deliveries of the gun – though it was not adopted as a standard weapon for the Army – and it seemed particularly suitable for the Marines. Unfortunately it had an appeal to gangsters large and small, a fact much deplored by Thompson, who was now a general. It became a familiar weapon to those who saw it used in action, real or simulated, by both police and criminals, in gangster films. As the average cinema audience, worldwide, was 230 million per week, the tommy gun became one of the best-known weapons in the world. In 1938, seventeen years after the first model had been perfected, it was officially adopted by the US Army.

In spite of its drawbacks of weight, cumbersome shape and heavy appetite for ammunition, the tommy gun was used extensively throughout the Second World War. It also appeared in subsequent conflicts, such as the Korean War and Vietnam War, though not in the hands of US troops, who by that time had different weapons. During the Second World War, as seen earlier, the tommy gun was phased out of the US Army (the last models were made in 1943) and its place taken by the M1 carbine. In the years immediately following 1945 sub-machine guns appeared to have reached the end of their useful life, their function being taken over by either assault rifles or pistols. It has been suggested that the day of the sub-machine gun is now finally over, but this appears to be a hasty judgement.

Although Britain had been very glad to receive large consignments of tommy guns from America during the early stages of the Second World War, the War Office took steps to design and develop a range of sub-machine guns of its own. First in the field was the Lanchester, a reliable weapon of 9-mm calibre, which could fire 600 rpm to 600 yards but at 9.65 lb was very heavy. Its chief customer was the Royal Navy. At that stage in the war tactical concepts based on sub-machine guns had not yet been evolved. However, by the end of 1941 there was a demand for a light, powerful, quick-firing SMG, mainly in the Middle East, and BSA evolved the Sten gun, Mark 1. This weighed a mere 7.2 lb and was cheap and easy to produce. It did not acquire its name from a Mr Sten – the S came from Colonel Shepherd, a director of BSA, the 't' from Mr Turpin, the chief designer, and the 'en' from the Enfield Arms Factory, where the gun first went into production. The Sten looked a crude and simple weapon, and was sometimes cursed by its users for stopping at critical moments, but overall it gave excellent service. The stock had been lightened by being constructed of skeletal metal and the gun had been made more sophisticated than the Lanchester by the addition of a flash hider. The models which succeeded the Mark 1 advanced to Mark 6, and each was probably more brutal-looking than the last. The Mark 2 dispensed with the skeletal stock altogether, replacing it by a metal rod with a plate on the end. The remainder of the gun looked like a discarded piece of motor machinery. However, the Mark 2 reduced the weight of the Sten gun to 6.65 lb. Although the appearance of the Sten was in such sharp contrast to that of most guns produced in British arms factories, famous for walnut butts and beautifully engraved stocks, it became immediately popular on account of its utilitarian character. There was no nonsense about the Sten, nothing much to go wrong or get damaged, and yet it was invariably useful and versatile. Resistance groups, to which the Sten was supplied during the Second World War, liked it because it could be concealed easily and could withstand rough handling. The Mark 6 (S) was fitted with a silencer, a feature which made it particularly suitable for special or subversive activities.

One of the more remarkable designs of the Second World War came into service soon after the Sten but was designed entirely separately. The Australian Owen gun, designed by Lieutenant E. Owen of the Australian Army, was heavier than was desirable, weighing 9.35 lb, but its other characteristics offset this. It fired at 700 rpm, and its muzzle velocity was 1375 feet per second. This latter was some 200 fps faster than the Sten and devastating by any standard. The thirty-two-round magazine projected *above* rather than below the gun (as was normal), but this, and the fact that it was so well balanced that it could be fired one-handedly, made it very popular with troops engaged in operations in jungle and scrub. The Owen appeared to have outlived its useful life by the end of the Second World War, but experienced a revival of fortunes when war began in Malaya in 1950. However, by the end of the 1950s the Australian Army was looking for a gun which would be lighter than the Owen but preserve its virtues. The answer came with the F1 SMG, a weapon which retained the vertical magazine but had a much shorter, 28-inch, length. Rates of fire were slightly down, 700 to 600, but weight was a full 2 lb less.

While the Sten was being developed, another light machine gun was being tested. This was the Patchett, named after its designer. As the Sten appeared to meet all known requirements adequately, development of the Patchett was not pressed, but it did go into production and a number were used by airborne units. In the post-war period several other British designs for an SMG were examined, but the only one to be approved, and adopted by the British Army, was the latest development of the Patchett. By 1953 this had acquired the name of the Sterling (it was manufactured at the Sterling factory in England) and its weight was down to 6 lb. Its rate of fire was the same as for the Sten (550 rpm).

In view of the keen interest that neutral countries had shown in producing guns either for or to be sold to future belligerents, it was no surprise to find that Sweden produced an SMG which combined the best features of the Finnish Suomi design ('Suomi' is Finnish for Finland) with characteristics of the British Sten. It was produced at the Carl Gustav factory, after which it was named. The Swiss also developed an SMG of their own and eventually evolved a very useful weapon in the post-war period. The Rexim-Favor was a neat and tidy gun with a weight of 7 lb, a rate of fire of 600 rpm, a length of 32 ins and a high muzzle velocity of 1300 fps.

Perhaps the unluckiest producers of SMGs during the Second World War were the Finns and the Italians. When the former had been over-run by the Russians, Suomi guns were sent to Russia for use by the Soviet Army, and when the latter were being kept in the war after they had tried to make a separate peace, their excellent Berettas were widely distributed amongst the German forces occupying Italy.

The description of weapons given above does not attempt to be comprehensive: it merely identifies weapons used in different countries and their firepower. There are many others; some, rather interestingly, are developments of guns which are in standard use in other countries. One such is the Finnish Valmet range. The Finns developed the Soviet AK-47 until finally it became the M82 Bullpup, a weapon with a 5.56 calibre which combined the qualities of the assault rifle with the portability of the SMG.

Pistols and revolvers were last discussed in Chapter 3, at the point where it seemed that their military function had been overtaken by the new range of automatic weapons, notably the sub-machine guns. However, whatever the developments in the SMG field, a personal weapon which could be carried in a belt holster was still likely to be needed. There had always been occasions when soldiers required a very small weapon, particularly if they were on clandestine missions. Although it was possible to conceal a Sten beneath an overcoat, this was not going to be much help if you planned to operate indoors or on a warm day.

Pistols have acquired a curious reputation from the writers of crime novels. Beautiful ladies extract small pearl-handled automatics from handbags and shoot their way out of difficulties with elegant nonchalance and astonishing accuracy. There is no doubt that a well-aimed shot, perhaps at the eyes or heart, could kill, but the chances of these miniature weapons shooting accurately or causing lethal wounds in the circumstances in which the heroine found herself do not seem high. As with stories of the Wild West, the pistol in fact and the pistol in fiction are two very different subjects. A typical example of the small pistol and one of the most bizarre weapons ever made was the stubby little Derringer, named after its American inventor, Henry Derringer. It had varying calibres, all enormous in relation to its size, a length usually under 5 inches, and, in the early stages at least, two barrels, one above the other, which gave it a two-shot capacity.

For the purposes of the First World War the six-chambered revolver, with a .38 or .45 calibre, was adequate. But the technique of manufacturing automatic weapons was already well advanced and in the early 1920s the resourceful John Browning, now in his late sixties, turned his mind to developing a new design for an automatic pistol. On this occasion he was co-operating once more with the Belgian Fabrique Nationale d'Armes de Guerre – the famous FN factory. He died before his latest design was perfected, but after his death more work was done on it by his staff, and eventually the final product, the GP (Grande Puissance) Mle. 35, became available in 1935. Although the USA had developed the M 1911, the GP was clearly a better gun and eventually was used worldwide. Unfortunately for the Allies, the Germans were able to use this design for their own armies in the Second World War, for they had captured the factory and plant when they over-ran Belgium in 1940. Although it was not unusual for both sides to be using each other's captured guns, it was unusual for them to be manufacturing the same weapon and using it as standard equipment.

Like all efficient pieces of machinery, the GP uses a simple design to achieve a complex purpose. The gun works on the short recoil principle and fires 9-mm rounds from its thirteen-round magazine, which is in the handle. It was clearly a suitable weapon for airborne troops, and in the sort of operations in which parachutists were likely to be engaged the fact that the Browning fired automatically only, rather than being able to alternate this with single shots, was a great disadvantage. It has now been modified to give it dual capacity.

In some ways the Browning GP was an inferior weapon to the Walther P.38. The Walther magazine held only eight rounds, but it could fire single-shot. This too worked on short recoil, and was adopted by the Wehrmacht in 1940. P.38s were manufactured in huge quantities and after 1945 found their way to many other countries, notably in the Far East. Otherwise, 1945 was a bad year for Walther, for the factory was in the Russian zone and was dismantled by the Soviets. However, the Walther family had not stayed to see that melancholy fate overtake their means of livelihood, and had prudently settled in the West. By the early 1950s they had set up a new factory at Ulm and in 1957 had the pleasure of seeing the P.38, later renumbered the P.1, adopted by the Bundeswehr. The P.1 has now been modified to make it the P.5.

Although these are the best-known names in the pistol world, they are by no means the only ones. The Swiss have equipped their very efficient army with the SIG Ordonnanzpistol 75; SIG is an abbreviation of Schweizerische Industrie Gesellschaft of Neuhausen-am-Rheinfall. Italian Berettas find much favour in Israel and Egypt as well as on the home market.

In the 1980s the public were intrigued to hear that the SAS were equipped with an unfamiliar name in the pistol world, the Heckler and Koch. There is nothing mass-produced about the H & K: it is the perfectionist's pistol. Certain models embody a number of refinements: one fires in three-round bursts, a fact which keeps it steady in the user's hand, conserves ammunition, yet is good enough to deal satisfactorily with the target.

The Russians do not appear to have developed any particular novel designs so far. Their Makarov is an efficient weapon, apparently developed from the GP design.

Over the years various methods have been tried to make pistol shooting more accurate. One, somewhat awkward, method was to fire with the right hand while resting the barrel on the crook of the left arm. Shooting accurately from the hip is not normally encountered outside the pages of fiction. Holding one arm out straight, while standing sideways to the target, has now been replaced by using both arms held out straight while facing forward and crouching slightly by bending the knees. The fact that the pistol user has to adopt these elaborate manoeuvres in order to make certain of putting his shots somewhere near the target explains why the pistol has now lost much of its credibility as a military weapon and is not to be preferred to the assault rifle.

It would be unfair to leave the pistol, limited though its contribution to firepower is, without some reference to its past achievements. No man ever inspected his weapon with such care as the man about to embark on a duel. The macabre eighteenth-century ritual of a duel, in which the contestants faced each other while a surgeon and a priest stood by to play their parts a few minutes later, depended not merely on the contestants' nerves but even more on the reliability of their weapons. Faulty pistols must have saved many a life. One of the reasons why duelling was abolished was that, as pistols became more accurate, there was an excellent chance of not merely one but both the contestants being killed. The pistol in fact put an end to a form of licensed murder which had begun in Roman times, though not much practised by the Romans themselves, and was originally a form of trial by battle. In medieval times many legal wrangles were settled by an official duel. Women were sometimes contestants. However, duels using weapons other than pistols have continued even in the present century: the Nazis made duelling legal again in 1936 and the Italian Fascist regime encouraged it.

An even smaller weapon than the revolver is the hand grenade. In its early stages the hand grenade or 'bomb' tended to be much larger. When first invented in the fifteenth century, it was a metal case filled with an explosive charge and a variety of uneven fragments; it was exploded by a fuse which was lit before it was thrown. Then, as now, once a grenade was ready for action there was no time for delay before it left the sender's hand.

In the nineteenth century in particular anarchists, revolutionaries and others of similar convictions occasionally attempted to change governments by assassinating the leading figure or figures. Several of these assassination attempts were successful: they were greatly helped by the fact that heads of state often drove in open carriages, showing themselves to a cheering populace. A bomb thrown at a man on horseback could easily miss and explode in the crowd on the far side; a carriage, on the other hand, formed a basket into which the bomb would fall and in which the explosion would be concentrated. Then, as now, bombs were also often left in suitcases, and were detected by the ticking of the alarm clock which had been set to trigger them off. No one, not even the most respected monarch, was safe from bombs thrown or placed by demented assassins. Alexander II of Russia, an enlightened autocrat who emancipated the serfs and earned the title of the 'Tsar Liberator', was none the less assassinated by an extremist's bomb in 1881. But the percentage of failures to the total was remarkable. Perhaps the most important and saddest assassination failure in history was that of Count Claus von Stauffenberg, who placed a bomb in Hitler's headquarters at Rastenburg in July 1944. It failed to kill the dictator, who took a medieval type of revenge on Stauffenberg and all whom he thought might be implicated. Two more eminent national leaders who were assassinated, though by rifle rather than bomb, were Presidents Abraham Lincoln and John Kennedy. President Reagan and Pope John Paul II have both survived one such attempt. Assassination of the leader of a nation at a critical point in that nation's history is probably the most effective use to which firepower can be put, if one is judging in terms of cost-effectiveness. The shooting of Mahatma Gandhi in 1948 by a Hindu fanatic undoubtedly caused avoidable bloodshed at the time of the Partition of India. An assassination which eventually caused the death of twenty million people was that of the Archduke Franz-Ferdinand of Austria, on 28 June 1914. He was shot with a pistol by a nationalist conspirator who was unaware that the Archduke, had he survived, would have aided rather than opposed the interests of minority groups in the Austro–Hungarian Empire. Once that fatal pistol shot rang out, it was taken as the signal for millions of others to follow. The world would never be the same again, and the remainder of the twentieth century would be spent trying to recreate the stability of the pre-1914 world.

The bomb and grenade are, therefore, potentially weapons of enormous and unpredictable power. The modern grenade is easily concealed. During the war in Malaya in the 1950s, General Gerald Templer transformed a campaign which the Malayan government appeared to be losing into one which it won decisively. In order to raise and maintain morale, Templer used to visit the kampongs (villages) and show himself to the local inhabitants. He became extremely popular, and security grew careless. On one of these visits an alert member of the crowd noticed a small boy, aged perhaps seven, pushing his way through the legs of people in the crowd in order to get a better view. That was natural enough. However, the spectator noticed a bulge in the rear pocket of the child's shorts, which could have been made by a tropical fruit – a papaya or durian perhaps. It was a very different kind of 'fruit': a Mills bomb, the famous 36 grenade, known ironically, on account of its appearance, as a pineapple. The small boy had been instructed to work his way to the front, pull out the pin and throw the grenade at the general's feet. It is, unfortunately, very difficult to take precautions against such conspiracies when the needs of public relations clash with those of security.

During the First World War, when opposing armies faced each other all along a 400-mile line of trenches, grenades were used in ever-increasing quantities. The two most famous were the British Mills bomb, a small, oval grenade filled with amatol, and the German stick bomb, which was shaped like a jamjar with a stick protruding from it. Both had delayed-action fuses, varying between four and seven seconds. Once the pin was removed, the bomb release was held merely by a lever. Holding this down, the soldier threw the grenade with a bowling action which was safer for him than a jerky throw which might cause the live grenade to slip out of his fingers and fall behind him. The target was usually an enemy trench or a machine gun post. On a few occasions grenades were picked up and thrown back, but extraordinary luck was needed for this to be possible. The Mills who gave his name to this extremely lethal weapon was Sir William Mills, who invented the device in 1915. The effective range of a Mills bomb was given as a circle of 25 yards radius, but on soft ground it could be less; there were, however, occasions when a grenade exploding on a hard surface such as concrete has killed at ranges over 100 yards.

Grenades seldom fail to explode. When this happens those nearby rarely forget the experience. On occasion a grenade has gone astray and the nearest soldier, to protect his comrades, has flung himself on top of it to absorb most of the blast. There can be few acts as courageous.

A different, but none the less memorable, experience with a grenade occurred to the members of the SAS squadron which captured the Omani rebel stronghold on the Jebel Akhdar in 1959. After the SAS men had climbed some 7000 feet to the summit of the mountain, during the night they encountered keen opposition from a well-armed rebel force which was holding excellent defensive positions. The enemy sniping was extremely accurate. Captain Walker, the troop commander, and CSM Hawkins had to deal with a sniper who was a mere 15 yards away but impregnably sited. Walker threw a phosphorus grenade followed by two 36s. The second 36 caught on a lip of rock and, rolling back into the gully where Walker and Hawkins lay, passed between them as they tried to press themselves into the solid rock. Then it went up, but by then it had settled in a fissure in the rock, from which the full force of the explosion was directed upwards. In consequence, neither man was hurt.

Hand grenades were soon used to contain materials other than lethal explosives. Phosphorus grenades can produce huge clouds of smoke immediately, enabling all types of tactical manoeuvres to be concealed. Coloured smoke is used for signalling, and has great value when ground troops are communicating to aircraft, perhaps in a period of wireless silence.

Grenades could be launched by rifle further and more accurately than by hand, although in the conditions of 1914–18 the hand-thrown grenade was easier and quicker to handle. During the Second World War rifle grenades were developed for anti-tank work. Firing at the front of a heavily armoured tank as it surged towards the gunner tended to be unsuccessful: his shots might miss altogether or, if they hit, be deflected by the armour. An anti-tank grenade would approach the tank on a different course, from which it could easily find a weak spot. In March 1945 an American infantryman aimed a grenade at a German Tiger tank. These formidable 65-tonners were usually more than a match for anything they encountered, but this one was unlucky. The grenade caught the edge of the turret and succeeded in exploding the ammunition inside: the tank was destroyed.

Grenades provided a simple method of making booby traps. In both world wars the Germans were particularly skilled in leaving behind interesting and innocuous-looking objects: tins of food, bottles of beer and magazines full of pictures of naked women were a few of the wide variety of eye-catching souvenirs which appeared to have been left behind in a hasty retreat. The unwary soldier who picked one up often disappeared in a sheet of flame. One of the simplest booby traps was made by putting a grenade in a lavatory cistern, so connected that when the lavatory was flushed the grenade was set off. Similarly, opening a door or treading on a loose floorboard could set off a grenade.

This type of inertia bomb was fully exploited by anti-personnel mines and anti-tank mines. These were set on exactly the same principle as the medieval booby traps described in Chapter 1. Anti-personnel (AP) mines were distributed so that the advancing soldier's foot would press down sufficiently hard to activate the charge. In order to prevent heavy losses among advancing infantry and tanks, trained soldiers were sent ahead, preferably in the dark, to probe the ground, locate each mine, gently separate it from the surrounding earth and then defuse it. Sometimes metal detectors were used; at others a bayonet was deemed adequate. This highly dangerous activity, usually performed by Engineers, was made even more so when live AP mines were laid one on top of the other. As the top one was lifted gently out of its position, it triggered off the one beneath.

Lieutenant (later Lieutenant Colonel) B. S. Jarvis was presented with the task of clearing a way through the minefields at Alamein in October 1942. He recalled:

> 'As we topped the crest the enemy opened up. The covering party rushed the post. Half of them were hit before they got there but they captured the chaps causing all the trouble. While they were doing this we probed for mines. Yes, there they were, our own, captured in June and relaid by the Boche. That was a good start. I found the far end of the field and placed my light [to mark the boundary]. The rest of the party got cracking and soon the gap was clear. It had worked perfectly.'[1]

When gaps had been made through minefields the cleared spaces were marked with white tape. The markings on these clear routes were noted by German aerial observers, who then informed their own gunners. As a result, tanks emerging from a cleared lane through a minefield were likely to be met by a barrage from the deadly German 88-mm guns.

R. F. Ellis was with the first wave of infantry to land on the Normandy beaches on 6 June 1944, and later recollected:

[1] Quoted in Philip Warner, *Alamein*. Kimber, London, 1979.

'I remember the three of us, Engineers, were called in to clear a path through the minefield surrounding the strongpoint. I recall one of our chaps being hit and being trapped by Spandau machine-gun fire with the bullets ricocheting off a small mound that I was lucky enough to fall behind. (In peacetime I suppose they call it fieldcraft.) Eventually I taped a path through the minefield. As I withdrew I distinctly remember the perspiration running from my head and the terrible dryness of mouth and throat – once again, was it fear? If it was I was too busy to think about it until afterwards. I must say I experienced more fear sitting in a foxhole when shells were falling, simply because I had nothing to do.

'I should point out that, because we worked as two and three-men teams, being attached to the South Lancaster and Suffolk regiments in the bridgehead, we saw action on all the Sword beach fronts, including mine-laying at night and night reconnaissance.'[1]

Although the years since 1945 are normally described as a period of peace, this statement appears to overlook the Korean War, the Malaya Emergency, the wars in the Middle East, the Vietnam War, the Falklands War, and worldwide insurgency and subversive activities. The last of these has led to a vast increase in the use of mines, booby traps and electronic warfare devices. A form of firepower much favoured by terrorists is the booby-trapped car, usually a stolen vehicle which has been packed with explosives. It is often triggered off by remote control, but sometimes is booby-trapped mechanically. A car full of explosive represents a massive quantity of firepower.

On the other hand, any counter-measure which reduces or cancels entirely the firepower of the other side is in itself a form of firepower. Now available to security forces such as police are a number of remote control vehicles which can pick up and remove suspicious objects or even take photographs. (This, of course, introduces the theme of electronic warfare, of which more will be said later.) Wheelbarrow is a small vehicle which resembles much less a wheelbarrow than a small crane on a tank-shaped chassis; it is powered by two 24-volt batteries.

[1] Quoted in Philip Warner, *D-Day Landings*. Kimber, London, 1980.

It is easy to control: a child can send it back and forth like a toy, and at military Open Days children are often allowed to do so. Under sterner conditions, Wheelbarrow can be instructed not merely to photograph the bomb from close quarters, but also to take essential steps to defuse it. Although Wheelbarrows frequently work to the point of no return, their ability to survive explosions has been remarkable.

A wide range of counter-measures now exists against the booby trap bomb. Some bombs cannot be defused with the time and resources available, and therefore are exploded under controlled conditions. One of the most effective methods uses a specially constructed nylon blanket. This, needless to say, is only effective against the smaller type of bomb. When monsters from the Second World War, 500–4000 lb in weight, are discovered, as they sometimes are by building workers and fishermen, the only practical step if they cannot be defused is to evacuate the local population until after the explosion.

Bombs which are too large to suppress on the site, but are small enough to be moved to a safe place, are taken away in a Total Containment Bomb Trailer. This has specially strengthened walls, so that if the bomb suddenly explodes when it is in transit the damage is minimized.

Very small bombs can do infinite damage if they are close enough to the right targets. In recent years many eminent people have been sent letter bombs, usually in the form of a bulky envelope or small parcel. Some of the early ones were very unsophisticated; they smelt of marzipan, betraying the presence of an explosive, and wires were visible immediately the package was opened. More recent letter bombs are altogether better constructed, and often employ materials other than metal. As some explosives may be triggered off by the proximity of X-ray apparatus, provision has to be made for the examination of suspicious objects at a safe distance. Similar types of X-ray equipment are used to detect explosives carried by people: handbags and suitcases are invariably given close scrutiny. This is important enough in public buildings, but vital in aircraft. An explosion on an aircraft may be used either to destroy it or to coerce a government by hijacking the machine and holding a country's citizens to ransom.

This is a new dimension in firepower, and one which is constantly altering. The early terrorist hijacks took the world by surprise; it seemed incredible that any group, however fanatical, would wish to jeopardize the lives of two or three hundred people, few of whom could have any influence on events, for a purpose with which the hijacked aircraft was totally unconnected. Any actions which placed the lives of innocent people in danger must surely be counter-productive. This, unfortunately, was not the way terrorists saw matters; their view was that governments which viewed with equanimity the fact that thousands of their citizens slaughtered each other on the roads each year were not likely to be unduly upset by the loss of a few hundred more, but the destruction of £4 million worth of aircraft, accompanied by a reluctance on the part of passengers to use that airline in future, might have political repercussions. Terrorists, who usually belong to very small groups, are unable to raise armies, air forces and navies to accomplish their purposes; instead they resort to a form of firepower which to a large extent relies upon the environment in which it is used. Thus an airliner, a passenger ship or even a train (such as was hijacked in the Netherlands with children aboard) multiplies the initial firepower of a grenade or a bullet by its own inherent vulnerability.

Regrettably, the only answer to the ruthlessness of the terrorist appears to be the adoption of an equally ruthless and callous attitude on the part of the threatened government, as in the Entebbe raid carried out by Israeli Commandos in 1976. Appeasement is rarely a successful policy; failures range from the Saxon attempts to buy off Danish raiders by paying Danegeld in 1008 to attempts to satisfy dictators such as Hitler, who was allowed to dismember, then occupy, Czechoslovakia in 1938. If airliners are hijacked again in the future, attempts will undoubtedly be made to repossess them by negotiation or by physical force; whatever the consequences, no government is likely to give in to terrorist demands without some form of counter-measures.

But, and it is a very large but, it depends on how the stakes can be raised by the terrorist. It is not impossible that one day a terrorist group will obtain possession of a nuclear weapon. It will be a brave government which will be prepared to see one of its principal cities demolished rather than assist the aims of a terrorist group operating for a cause in which Westerners have no interest. Mounting such a threat would be unproductive in the long run, but terrorists are not interested in long runs, only in short-term effects. This might seem to others to be short-sighted, but they do not think like the terrorists. Fortunately, since all governments, of whatever political creed, are well aware of the possibility of such a calamity, it is less than likely to happen.

A recent reminder that firepower is related less to the size of the explosive than to the importance of the target occurred in the Brighton hotel bombing in 1984, when an attempt was made on the life of the British Prime Minister by concealing a bomb behind a panel in a room which she would be occupying several weeks later. This brought to the notice of the public the fact that timing devices could now be set for explosions to take place months later. Once again it looked as if the most important aspect of firepower in the future would be the electronic device.

Once, bombs were thrown at national leaders by desperadoes who approached as near as they could and risked their lives in consequence. Now such bombs are planted in conditions of safety, in rooms or under cars; the conspirator does not risk his or her life by approaching close to the victim: he plants the bomb and waits for the victim to walk unsuspectingly into danger.

6

Bigger and Better Guns

Chapter 2 left artillery in the sixteenth century, when the performance of guns and gunpowder was becoming a matter of science rather than luck. As guns grew in size, weight and length, it seemed particularly appropriate to use them in ships. Henry VIII was the great pioneer of naval gunnery: he had the money and determination to ensure that his fleet was one of the best armed in the world. His flagship the *Great Harry* carried four 60-pounder cannon, twice that number of demi-cannon (32-pounders) and a quantity of lesser artillery. This heavy naval armament, and the fact that sailors had to fire the guns as well as sail the ships, gave an advantage over the Spanish fleet and led to the defeat of the Great Armada in 1588 during the reign of his daughter Elizabeth.

But, although the Spaniards lost the Armada through inferior gunnery aided by inefficient seamanship and bad weather, they made good use of gunpowder on other occasions. In various parts of the world, notably in South America, it gave them an enormous advantage. However, although it was generally recognized that cannon and bombardment could be used very effectively against castles, forts and ships, no attempt was made to integrate them into battlefield tactics. Pistols and rifles had obvious applications: heavier guns seemed far too slow, unwieldy and troublesome in relation to the damage they might cause. Gustavus Adolphus of Sweden (1594–1632) changed that attitude completely by introducing light artillery pieces which fired 3-lb shot with accuracy; field artillery had been born. There was, however, much to be done before large guns would be fully mobile. At first it was not realized that moving the guns should be a function of the gunners and not of anyone who could be pressed into service. One result of hiring or press-ganging undedicated drivers into service was that they were likely to depart hastily, leaving the guns stranded, if enemy fire became embarrassingly close. Once this problem had been solved, the future of field artillery seemed assured.

Marlborough was the first British general to use mobile artillery successfully, and did so at the Battle of Blenheim on 13 August 1704. He took a variety of guns, large and small, to the battlefield, but the most effective were probably the 3-pounders because of their high degree of mobility. Nevertheless, every gun needed a horse team to drag it around, the horses needed forage, the guns needed ammunition, and the supply train grew longer and longer. The difficulties involved in moving guns over the poor roads of that time imposed a limit on the size which could be used. Only after many years' experience, combined with a general improvement in the condition of roads, could the bigger pieces be brought to the battlefield. Nevertheless, after a slow beginning the science of gunnery began to develop rapidly. The British Army's recognition of the importance of this arm was shown by the foundation of the Royal Regiment of Artillery in 1722. Another stride forward was the foundation in 1741, under distinguished patronage, of an academy for training artillery officers. When first established at Woolwich it was entitled the Royal Academy, but twenty-three years later the name was changed to the Royal Military Academy. The title is still in use, although the RMA has greatly changed, trains officers for all corps and regiments of the service, and is now at Sandhurst, Berkshire, where it has amalgamated with the Royal Military College.

At the Battle of Dettingen in 1743, the last occasion on which an English king (George II) personally led his troops into action, artillery was used enthusiastically by both sides. This account of the battle was written by a man named S. Davies who took part in it:

> Our battel lasted 5 ours, the first they played upon our baggage for about two ours with their cannon, the ball was from 3 lb to 12 lbs each. We stayed there till the balls came flying all round us. We see first a horse with baggage fall close to us, then seven horses fell apace, and then I began to stare about me as the balls came whistling about my ears.

Davies's remark 'began to stare about me' signified much more than apprehensive curiosity. With a cannon about 1000 yards distant from its target, it was possible to observe a ball leave the gun and set off on its way. An alert person in the target area kept a sharp lookout for any which appeared to be coming in his direction. Owing to their low muzzle velocity, he would have time, if he was not otherwise engaged, to move out of the line of approach.

All cannon at this time were muzzle loaders, fired by igniting the charge through a hole at the rear end. Cannon usually fired ball, grapeshot or canister 'point blank', in other words directly at the target; howitzers, which had much larger calibres (up to 10 ins) sent heavy cannon balls and incendiary material on a high trajectory which enabled them to land behind the enemy defences. The word is derived from the German *Hauvitze*, which in turn comes from the Czech *houfice*, meaning a catapult. Too heavy to be moved far or fast, howitzers were mainly used in sieges. In 1784 Lieutenant Henry Shrapnel, an officer in the Royal Artillery who spent much of his time in private research, invented the form of shot which eventually bore his name. At first this was a ball filled with bullets; it exploded a time fuse as it reached the enemy line. Subsequently the ball became a cylinder. However, improved explosives made the secondary charge unnecessary, as the shell casing itself became shrapnel as it disintegrated. Nevertheless, although the original design had become obsolete, fragments of high-explosive shell continued to be called shrapnel.

These developments were not without their problems. Sometimes the propellant burst inside the gun; sometimes the shell went off-course, with disastrous results. The problem with artillery, then as now, was that when a shell went to the wrong target it caused exceptional damage. In earlier times accidents were caused by the unreliability of the charges; in later times they have often been the result of what is known as 'human error'. On one occasion in the Second World War a group of senior officers gathered to witness the potential of a new weapon. Unfortunately, owing to an incorrect calculation, they did more than witness it: they experienced it. A stray shell landed in the middle of the group, causing numerous casualties. In peacetime accidents have tended to be less disastrous, owing to stringent safety precautions and non-explosive shells. Even so, there have been some occasions in the last few years which artillerymen would prefer to forget, such as a shell from a nearby range landing in the garden of the divisional commander, and the spire of a medieval church being knocked off by a gunner who had mistaken it for a target.

In the early years of the nineteenth century a new concept of the part artillery might play in war was discovered by chance at the siege of San Sebastian during the Peninsular War in Spain. As the British infantry advanced, the gunners first shelled the ground immediately in front of them; when that ground was about to be taken, the gunners lifted their fire to a point further ahead. From this developed the creeping barrage, by which artillery time their shelling zone to advance at a fixed rate per minute, perhaps 25 yards. When the advance goes as planned, and the shelling is properly co-ordinated, the results are very satisfactory. However, in the First World War there were many occasions on which the attacking infantry were unable to make the progress they expected owing to wire, enemy machine gunners or mud – often all three. On those all too numerous occasions the creeping barrage, which was meant to protect the advancing infantry, was soon ranging far ahead of them. Immediately it had passed, the enemy gunners emerged from their shelters, where all but a few had been taking cover, and proceeded to blast the oncoming infantry as it advanced over open ground.

Gunnery was made infinitely more efficient in the mid-nineteenth century by the invention of first the percussion tube and then the friction tube. The former ignited the propellent charge by means of a fixed hammer, set in motion by a pull on an attached lanyard. The latter consisted of an iron bar with a tough surface which moved rapidly through detonating powder; this too was controlled by a lanyard.

Until the middle of the nineteenth century the design of rifles was in advance of that of bigger guns, but in the second half of the century big-gun development began to make impressive strides to redress the balance. For the Crimean War a number of smooth-bore guns were converted to the rifled principle. The first example of this, the Lancaster gun, did not embody the true rifling principle but instead had an oval projectile moving along a barrel which was bored in oval form; it turned the shell once in 360 ins. Although this amount of twist did not impart a fast spin, it was enough to make Lancaster guns a great improvement on their smooth-bore predecessors; they achieved impressive results at the siege of Sebastopol in 1855.

However, the previous year had brought another new design of gun, the invention of Mr William George Armstrong, a civil engineer from Newcastle-on-Tyne. Armstrong was a lawyer turned engineer and industrialist, and types of gun were only one of his inventions. He made a close study of the liability of cannon to burst when the propellent charge was increased beyond a certain point; this was especially likely if the bore had been adjusted to turn the shot. As an engineer he was aware that, if the tension on the outside of a tube can be strengthened, the power of the whole tube to resist an internal explosion is greatly increased. He therefore built up gun barrels by shrinking iron plates on to an inner steel tube. Inside this much strengthened barrel the outgoing shell was now twisted by a number of shallow grooves. The breech was closed by what at first was called a vent piece but is now known as a breech block.

Nevertheless, these experiments fell some way short of complete success. The moment of truth came when some of the new inventions were tried out in the American Civil War of 1861–5. Most of the trouble occurred in the breech blocks, which often proved too delicate and complicated for the rigours of war. Similar problems were encountered with other guns. As a result there was a temporary reversion to muzzle loading – a retrograde step, but at least the gunners were less likely to be killed by their own guns. The usual acrimonious debate broke out between those who favoured the new inventions and those who thought it would be better to rely on the type of gun which had helped win the Battle of Waterloo. The latter school of thought was much encouraged when several European armies experienced trouble with their breech loaders. Krupp, considered by many to be almost infallible, suffered as badly as anyone with his. There were many failures in the Franco–Prussian War of 1870–1, though not enough to please the French.

But in the course of ten years most of the problems of the breech loader had been solved. A longer projectile, a device for preventing the escape of gas and a new form of powder all played their part. In 1878 it was realized that guns could be built in two parts, which made them easier to transport. A 2.5-inch gun constructed in this manner earned the name 'screw gun': the two halves of its barrel were screwed together when it was needed for action, and it proved a highly successful weapon in mountain warfare.

Towards the end of the century the slow rate of fire of many guns was beginning to cause concern, mainly in connection with coastal defence. There were two main problems: the breech mechanism needed to be less complicated and therefore quicker to operate; and the recoil had to be absorbed more quickly if rapid, controlled fire was to be possible. The solution to the second of these problems was found in an improved gun carriage.

The South African War of 1899–1902 taught artillery officers and designers many lessons through hard experience. The first was in tactics. Up until then guns had been brought up to a suitable point near the line and then fired at a visible enemy. Troops in fortifications, or other places of concealment, could be dealt with by howitzers. Unfortunately for the British artillery, the Boers did not fight the war as was confidently expected. Their artillery exceeded the British in range and power (it included some excellent guns from Krupp) and they had a similar superiority in small arms. Furthermore, they did not wait to be shot at: they delivered their own barrages from concealed positions and moved on hastily before an effective reply could be made. And when that reply came it was anything but effective. The field guns with which the British artillery began the war were 12-pounders which had been rebored to make them take a 15-lb shell. The process of reboring had weakened the barrels so much that the 15-lb shell could only be used with reduced charges. As the unfortunate British gunners struggled to bring these obsolete monsters into action, they were sniped at by Boer marksmen who were not merely adept at fieldcraft but were also using smokeless powder.

Previously, firing had been done over what is called 'open sights', that is with the gun pointing directly at the visible enemy. But now, alas, the enemy was not visible: he was an expert at concealing himself in trees or behind ridges. The Boer War demanded a revolution in attitudes to camouflage: the British Army had in fact only recently abandoned its habit of going into action wearing red. (Red was attractive to recruits for it caught the eyes of the girls; unfortunately it also caught the eyes of the enemy marksmen.) But camouflage and concealment were not the only skills to be learnt in this very difficult war. The most important was what came to be called 'indirect fire'. Indirect fire was in fact directed fire: when, for example, the enemy was on the far side of a ridge, the artillerymen posted observers on top of that ridge or, if that were not possible, on some point which could enable them to report on the accuracy of their own shellfire.

But to make proper use of these forward observation points required special qualities in the observers, not least of which was some knowledge of mathematics. The forward observation officer (FOO), as he came to be known, had to be sufficiently agile to climb to the most suitable observation point, to be able to estimate from a puff of smoke how near a shell fell to the intended target, and, not least, to have the appropriate skill to convey his findings to the men on the gunsite. Modern gunnery practice had at last been born, but it had been a difficult labour. Indolence masquerading as respect for tradition had helped to retain many obsolete and dangerous practices. Not least had been that of polishing guns until they shone magnificently in the sunlight, visible to all and especially to the enemy. The official decree which said that in future guns used in campaigns should be painted khaki seemed to many of the older officers to be akin to dressing the Venus de Milo in a coat and skirt.

The Boer War had shown how necessary efficient, quick-firing guns would be in any future conflict, and both Britain and Germany embarked on intensive research in this field. Unknown to them, the French had already solved this problem. Their 75-mm gun, which had been produced in 1896, could fire at thirty rounds a minute. Not surprisingly the French, who considered that they might be engaged in a war against Britain in the near future and were certain that they would soon find themselves fighting Germany once more, tried to keep the design a close secret; but it did not remain so for long, and soon a British version was in existence. The only problem for the War Office was whether the new gun should be a 13-pounder or an 18-pounder. In the event, both were manufactured, but during the First World War the 18-pounder had a hundred times more use than its smaller brother.

At the beginning of the present century, as all armies pressed ahead with programmes of rearmament, problems of mass production began to show themselves. Faulty components were delivered to the armies, who complained bitterly but with little result. Not the least of the complaints from aggrieved battery commanders was that some of the barrels were not even straight. 'How would one hope to shoot straight with a crooked gun?' they asked in tones of increasing bitterness. 'With high-velocity shells,' they were told. To their surprise the answer turned out to be true. The shell straightened the barrel as it passed through. But other faults were not so easily remedied.

A companion to these quick firers was now being developed in the howitzer range; this was the 4.5, but it did not finish its trials until 1908. Once the gun had proved itself it was soon followed by a 9.2-inch, a 12-inch and a 15-inch. As problems of transport were gradually overcome, a new range of even heavier guns put in its appearance. Few at that time realized that war with Germany, if it ever came, would be a combination of massive bombardments and suicidal infantry attacks.

In the early years of the twentieth century concepts of the use and possibilities of artillery were required to take a giant, imaginative stride forward. Ranges had doubled in five years, fuses had improved greatly and explosives were more stable and powerful. All these factors demanded that the artilleryman, especially the officer in the field, should be skilled and well practised. The latter requirement created a need for artillery ranges, and an unavoidable aspect of range training was that it required huge areas of land. Parts of Wales, Wiltshire and Devon were acquired by the War Office, a change of ownership by no means invariably popular with the local residents. However protests, then as now, seemed to be of little avail. Recently, however, army ranges have found unexpected friends among conservationists. Because the army occupied large tracts of land which were never ploughed except by shells, traditional but increasingly rare plants continued to flourish on them. The disturbance caused by shells was as nothing compared with the havoc which otherwise would have been wrought by plough and chemical defoliant. Nature makes strange friends.

Directing fire – or people – by means of the clock code, that is projecting an imaginary clock face on to the scenery ahead and indicating a position by its proximity to one of the hypothetical clock's numbers, now seems so natural a practice that one tends to overlook the fact that it must once have been someone's new and intelligent idea. It dates from the early twentieth century. Unfortunately the name of that progressive thinker is not known; there should certainly be a memorial to him, anonymous though he be – perhaps a pillar with the hands at five to one?

One of the most perilous means of target spotting was to ascend in a balloon and report, by trailing telephone, from that position. Since the first ascent in 1783, the balloon had frequently been used for observation of the battlefield. In 1794, when the French were fighting the Austrians, they sent up an observation balloon with an intrepid captain as its occupant. This feat was quickly emulated by other nations. Kitchener went up in a French artillery spotter's balloon in 1871 when still a British Army cadet, and got into serious trouble for doing so: he was reprimanded and he also caught pneumonia. In the nineteenth century the hazards to a balloonist were already considerable, but in subsequent years the development of the military aircraft increased them many times. The presence of observer balloons and enemy aircraft in the sky gave a fillip to another aspect of artillery, the anti-aircraft gun. The year 1909 saw the demonstration in Frankfurt of guns manufactured by Krupp which had been specially designed for use against balloons. Being static, balloons were vulnerable, but a moving aircraft was a very different matter. The problem was never satisfactorily solved, but the 3-inch anti-aircraft (AA) gun which went into service with the British Army just in time for the First World War created an impressive and noisy spectacle, even if its effect on aircraft was almost negligible. Until recent years, anti-aircraft guns have been respected more for their rôle in helping civilian morale than in reducing the amount of enemy air activity. They do, of course, create an alarming barrage in the air, but as a defensive curtain this is much less effective than a barrage made of balloons – the object which originally brought the AA gun into being. In the phonetic alphabet of the day A was Ack. Although the Ack has now become Alpha in conformity with NATO standardization, the term Ack-Ack shows no sign of dying.

The First World War was not merely infinitely more bloody and lengthy than had been predicted; it also took a different form. Mobility had been the order of the day in the Boer War except for the early sieges; in the world war which followed, the Western Front, which was the vital sector, settled into a gruelling, tenacious battle for a few miles of devastated land. The war had begun with Germany thrusting deeply into Belgium and France. When that attack was checked, as it was at the Battle of the Marne, both sides engaged in what was known as 'the dash for the sea' but was in fact an attempt to construct a 400-mile-long defence before it could be outflanked by the enemy. Inevitably, the rush to form tenable lines meant that certain points were more vulnerable than others. Periodically desperate attempts were made by either side to break through the opposing defences and exploit the gap created. This policy, though clearly costly and unsuccessful, was adopted by both sides. The British made heroic efforts to penetrate at Loos, at Festubert, on the Somme and at Passchendaele; the French, who had great faith in the ability of their artillery to smash a hole through the German lines, broke through at Artois, gained three miles of ground, lost them again, and found themselves back on their original line, having lost four hundred thousand men. That was in May 1915, and when they tried again in Champagne in December of the same year the results were equally unsatisfactory. In April 1917 the French tried again, this time near Laon, and lost a hundred thousand men. This failure, following the enormous losses they had suffered in the heroic defence of Verdun in the previous year, brought the French Army to the point of mutiny and imminent collapse. It was rallied just in time by Pétain, but the effects of the French reverses and losses would linger not merely through the rest of that war but also into the beginning of the next, over twenty years later.

The Germans were equally hidebound in their tactics. They had begun the war with a daring but complicated encirclement plan devised by Graf von Schlieffen which made heavy demands on their logistic support. When they were checked on the Marne they fell back on to satisfactory tactical positions – all of which had been studied by a pre-war army of spies masquerading as railwaymen, hotel workers and tourists. One of those positions was the low range of hills around Ypres, which would be the theatre for some of the greatest gunnery battles of the war. But the Germans' efforts to break through the Allied lines in that sector, for which they used gas, proved unsuccessful. Although there were weaker sectors, in many the trench line was so well constructed that it remained virtually static for the rest of the war. Nevertheless, the Germans were as stubborn as the Allies over their tactical approach. Falkenhayn, the German commander, believed that if Verdun were to be battered relentlessly for twelve hours along a 20-mile front the French defence would collapse. His estimate proved to be very far from the truth. For four dogged months the Germans pounded away at the French defence; but this was more than a twentieth-century battle, it was the resolution of centuries of feuding between Teuton and Celt, and it was fought with a desperate hatred nurtured on something stronger than mere nationalism. The Germans launched continuous suicidal attacks, but still Pétain could say: '*Ils ne passeront pas.*'

In the spring of 1918, when the French were still trying to recover from the effects of heavy losses, a mutiny and unrest at home, and when the British were still preoccupied in recovering from the heavy losses incurred in the Passchendaele campaign, the Germans made one last desperate attack. On 21 March Ludendorff, who was now in command, opened an offensive between Arras and St Quentin, with the intention of reaching Amiens. A tremendous opening barrage was followed by mass infantry attacks, and the result was a breakthrough which took the Germans 40 miles onward within a fortnight. Ludendorff now delivered his secret blow on the Lys, aiming at the Channel ports. But the British in this sector held fast and the line was saved. In May he struck against the French, captured Soissons and advanced towards the Marne. Here his troops were once again brought to a halt.

But they were not yet finished. After nearly four years of war the Germans still had the strength to strike once more against the reeling Allies. This time the thrust was to Compiègne, but again it was held after a few miles. Finally, in July, Ludendorff made a desperate last throw at Rheims; a breakthrough here would open a way to Paris. But on 18 July it was all virtually over. The sequence of German offensives at last stopped and now it was the Allies' turn.

The British began the long fight back in the Amiens sector, where it had all begun. The Americans followed by attacking at Saint-Mihiel, and the French began an offensive in the central sector. These were the major campaigns. Each had begun as a battle, but each lengthened into a campaign. In the intervals between the offensives, all of which began with devastating cannonades and continued with mass attacks and constant barrages, there were what were officially described as 'periods of quiet'. Periods of quiet lasted for weeks and months, during which both sides shelled steadily according to the amount of ammunition available. At intervals local attacks were carried out under the orders of commanders who wished to straighten untidy sectors of the line by capturing territory from the enemy. Through it all, the guns of both sides pounded each other's front line trenches, communications and rear areas. There was never any respite. The 'morning hate' in which both sides put over a few dozen shells soon after dawn became such a routine that it seemed part of normal existence.

The Germans had begun the war with marked faith in the ability of heavy guns to destroy everything which stood in their path. This view appeared to be confirmed when their 42-cm howitzers smashed their way through the forts at Liège, Namur and Maubeuge. But for trench warfare a different type of gun and different tactics were required, although the Germans never completely lost faith in the ability of the heavy gun to win the war. In 1918 the Paris gun shelled that city from a range of 67 miles. The immediate result of shells landing in the French capital at fifteen-minute intervals was gratifying to the Germans, but the material achievements were less impressive than the psychological effects. At the end of the first day, twenty-one shells had killed fifteen people but no important targets had been hit. On the following day a further twenty-two shells killed eleven people, and on the day after that six shells killed one person only. Later results were more impressive: a shell which landed on a church caused the roof to collapse and kill eighty-six members of the congregation. Eventually the number killed mounted to 256, with a total of 620 wounded, but Paris was neither disrupted nor demoralized by the experience.

The Allies never managed to locate and destroy the Paris gun, but they did subsequently learn some details of its construction. It was actually a 38-cm naval gun which had been fitted with a 21-cm lining. The lining was extended and fitted into a smooth-bore tube, giving the weapon a 130-foot barrel. As may be expected, the wear and tear on a gun of this size gave it a short life, and after firing sixty shells the barrel had to be returned to Krupp for reboring. It returned with a 24-cm lining. An extraordinary mystery of the Paris gun and its various replacements was that, apart from a single, unused gun platform, no trace of them was ever found. It was speculated that the barrels were cut up for scrap, but if that had happened those who had been concerned in their disposal must have been sworn to secrecy, and the secret well kept. Rumour said that they had been hidden, waiting for 'the next time', but they never reappeared. The Paris monster was nicknamed 'Big Bertha' in honour (*sic*) of Bertha Krupp. The gun weighed 90 tons, but apparently the lady was flattered by her name being linked with it.

Krupp guns have already been mentioned several times in this book, and the contribution of that remarkable family to international firepower merits some examination here. The Krupp dynasty began in 1587, when Arndt Krupp bought large tracts of land in Essen from owners who, on account of an approaching plague, were very happy to release them for modest sums. The family was already prosperous when it began steel manufacturing in 1787, but from that moment it grew rapidly into an armaments empire. Guns were, of course, only a part of the Krupp production; they also built railways, ships, submarines, hotels, banks and cement works. Krupp did not merely arm Germany: it supplied weapons to dozens of other countries too. With this entrepreneurial skill went considerable eccentricities. Albert Krupp (1812–87) believed in the therapeutic value of the smell of horse manure and built his study over a stable from which he could inhale its benefits; he was also so terrified of fire that he would never live in anything built of wood. His son Frederick (1854–1902) was even more gifted than his father in the art of inventing, manufacturing and trading armaments. His downfall came when his homosexuality became public knowledge. He had married earlier and had two daughters, but the public scandal was too much for him and he committed suicide. His eldest daughter Bertha succeeded to the Krupp empire, and in the opinion of the Kaiser needed to be married to a suitable husband as soon as possible. He chose for her a Prussian named Gustav von Bohlen, who, on marriage, had his name changed to Krupp. His hobby was reading train timetables. It was Gustav who decided to name the bigger guns after his wife, and he also financed and organized the rearmament of Germany after the First World War. He financed the Nazis, supported Hitler and persecuted Jews. By 1940, although only fifty years old, he was suffering from advanced senile decay (he lived for ten more years) and Hitler insisted that he should hand over control to his son Alfred.

Alfred had already shown signs of the family talent. As well as encouraging and financing the notorious Waffen SS, he had supervised the production of a new range of guns of which the best known was the 88-mm, an anti-aircraft gun which functioned in many other roles later; it was especially famed as a tank killer in the Second World War. His ruthlessness towards the slave labour he used in his factories later brought him to trial as a war criminal. At Nuremberg he was stripped of his holdings and sentenced to twelve years in prison, but less than four years later he was pardoned and given all his property back. The firm prospered as never before; Alfred's personal fortune was so vast as to be immeasurable.

But nemesis awaited. Recession hit the firm in 1966, and a year later Alfred was found dead. His son, Arndt, had no wish to take over the Krupp empire, so it became a public company. Young Arndt was granted a pension in return for relinquishing all claims to Krupp; he was not rich by his family's standards, but the sum of a million Deutschmarks per annum enabled him to live in reasonable comfort.

At the beginning of all wars military philosophers usually agree on one point at least – that this one will be a short one. The British are particularly prone to optimism in this respect and at the beginning of the Boer War and the First and Second World Wars there were few who did not believe in a reasonable chance of it being 'all over by Christmas'. In the First World War, the Germans who attempted to implement the Schlieffen Plan were expecting to be in Paris by the end of the first month; in the Second they were slightly less optimistic, in spite of their plans for a Blitzkrieg when the time seemed auspicious. They never reached Paris at all in the First World War, but were there after nine months of war and a five-week campaign in 1940.

When the war which had begun in August 1914 showed no signs of ending quickly there was widespread surprise and much rethinking. One man, of course, had not thought it would all be over by Christmas, but instead had predicted a duration of at least three years. That was Earl Kitchener, a field marshal with successful, though lengthy, campaigns in the Sudan and South Africa behind him. Disregarding the incredulity of his colleagues, Kitchener set about raising an army of over a million men. However, when the war was obviously going to follow the course he had prophesied, Kitchener was blamed for the munitions shortage.

A munitions crisis was inevitable – for everyone. For the early stages, the majority of the guns were light field pieces. The Germans had used heavy howitzers to batter their way into Liège and Namur, but they had no more conception than anyone else of the vast numbers of big guns which would eventually be brought into use. And it was not merely the number of shells required which would produce problems, but the variety. The British Expeditionary Force had set off to France with 486 guns. Four years later it possessed 6437. At the Battle of the Somme in 1916 there was one heavy gun every 57 yards; in the opening stages of the Battle of Passchendaele a year later there was one every 20 yards. In some sectors guns were drawn up with their wheels almost touching.

'Now thrive the armourers,' was the immortal phrase put into the mouth of Henry V by Shakespeare. It certainly applied to World War I.

As guns increased in numbers, so shells increased in power and sophistication: there were armour-piercing shells, shrapnel shells, gas shells, smoke shells and incendiary shells. Armaments manufacturers regard their rivals' successes with equanimity. When one firm produces an armour-piercing shell, another produces a new and tougher type of armour to resist it. When one produces a battleship, another produces a submarine, a third produces a depth charge and a fourth an acoustic mine. Often they supply each other with components. In 1902 Vickers adopted a Krupp fuse for its shells: Vickers shells were accordingly stamped KPZ (Krupp's Patent Fuse). A charge of one shilling and threepence was levied. This would be eight pence today if directly converted, but would in real terms be more than one pound sterling. At the end of the First World War Vickers settled the bill for the previous four years – during which, of course, there had been no payment. The calculation could only be made by reference to the German casualty returns, which provided information on the deaths caused by British artillery. It was not an exact figure, but it amounted to the equivalent of several million pounds. It was slightly embarrassing for Krupp to have made a profit from the number of German soldiers killed with the assistance of one of their fuses but, as the French could put it, *'C'est la guerre.'*

Although Krupp is probably the best known and most awe-inspiring name in the armaments field, there are plenty of others of no inconsiderable stature. The French family of Schneider, who had built the first locomotive and the first steamboat in France, created an industrial empire in the nineteenth century, first in the Ardennes and then at Le Creusot. Schneider was what is now called 'fully diversified'; they manufactured everything from turbines to tanks.

At the beginning of the First World War the Austro–Hungarian Empire still existed: by the end its disintegration had begun, and it ceased to exist when the peace treaties were drawn up. In its day, that Empire had ruled a vast, ill-assorted collection of peoples; among them were Bohemians, occupying the western part of what is now Czechoslovakia. At Plzen (Pilsen), in 1819, had been born Emil von Skoda, founder of the great Skoda armaments works. Emil had studied engineering in Germany before becoming chief engineer in a small factory in his home town. He soon bought out his employers and then expanded the business at astonishing speed. When he started producing arms towards the end of the century it was clear that his factory would become a giant in that field. His 30.5-cm howitzers earned the admiration of all, particularly of the Germans, who gladly included them in their forces bombarding Namur in 1914. From Namur, in company with the Krupp Berthas, they went on to Maubeuge. Two of these Skoda monsters may be seen in the military museum at Vienna and there is no mistaking their latent power.

The Skoda factory continued to produce arms for the German war machine in the Second World War. It had been acquired by Hitler because the Munich Agreement of 1938 permitted the partition of Czechoslovakia. He did not, of course, take over the country immediately; he was content to begin the process by which an admirable democratic state was made to disintegrate. The final stage took place in March 1939, when he established a German Protectorate of what was left and which, naturally, included the Skoda works.

Massive though the output of all these First World War arms manufacturers had appeared at the time, it was easily eclipsed twenty years later. By the end of the Second World War, arms were being manufactured all over the world on an unprecedented scale: in the USA alone it reached totals which a few years before would have been thought impossible.

But guns were never an unbroken success story. Accidents occurred with such regularity in the First World War that there were constant rumours of sabotage. Until inspection was rigorous, shells were liable to fall between the two extremes of either being too powerful for the barrel and bursting it, or being too weak to do adequate damage to the target. Many failed to explode at all, a most welcome fact to those in the target zone. The French had more trouble than anyone, until the war had been going on long enough for them to reorganize themselves. Nevertheless, the 75-mm was one of the best guns in the war, and in the final stages was used extensively by the US Expeditionary Force.

While the howitzer was getting to work on the bigger targets, the everyday task of demolishing installations in and immediately behind the trench lines was accomplished very efficiently by mortars. The original trench mortar was a German invention, as simple as it was diabolically effective. Named the Minenwerfer, it consisted of a steel tube 2 feet long and of 90-mm calibre, which was held in an easily portable steel frame. Within this the mortar could be aimed at anything within 600 yards' range. A bomb was slid down the muzzle and, when fired, went wobbling through the air in the direction of the target. On arrival, a special fuse ensured that however it fell it would explode. The 'Moaning Minnie', as it quickly came to be called by the British, was one of the most hated of the German weapons, but that did not prevent its being copied by British manufacturers for British use. Improved versions, with a longer range, soon put in an appearance. The fact that both sides now seemed to be committed to fighting the war from trenches gave this new weapon an added importance. Remarkably, the most efficient design was that of Wilfred Stokes, managing director of a firm which in the first half of this century was known to almost every owner of a lawnmower – Ransomes. Ransomes produced many other pieces of machinery apart from lawnmowers, but probably nothing which became as well known as the Stokes mortar. This began as a 4-foot-long, 3-inch calibre, smooth-bore muzzle loader which had the advantage of weighing only 36 lb, but soon acquired a younger, but bigger, brother, the 4-inch mortar. Both were remarkably reliable and accurate, and both were sufficiently adaptable to be used as anti-aircraft guns. In this sense, as a portable, powerful, trouble-free, adaptable infantry weapon, the mortar was a new invention; however it was also the successor to the clumsy bombards used in the very early days of gunpowder. And it was also first cousin to the rocket, whose ancestry went even further back in history than that of the mortar; the rocket, in abeyance in the First World War, would once more leap to prominence in the Second and in 1987 be classed, with lasers and a few other deadly mass killers, as 'the ultimate weapon'.

Since its development in the First World War the mortar has remained in favour as an essential infantry weapon. It gave tremendous service in the Second, when all the belligerents used mortars based on the design of the French Brandt 81-mm. The Brandt was simplicity itself. A 7-lb bomb was dropped down the muzzle, struck a firing pin and then landed on a target up to 3000 yards away. More powerful models were made, but these came closer to becoming small howitzers. The Russians, the Germans and the Japanese all used mortars with great skill and effect. The Russians adopted the French design to manufacture the 120-HM 38, which could accurately launch a 35-lb bomb to a distance of 6000 yards. Both Germans and Russians used each others' mortars if they captured them, and had no hesitation about simply copying design improvements into their own manufactures. The Japanese found the mortar a very useful weapon in the jungle. If they could goad their opponents into giving their position away, either by verbal taunts or chance shots, they would promptly saturate the target with well-aimed bombs. By the Second World War the Germans were calling their weapons Granatwerfer (grenade throwers), although they still used the term Handgranate for normal-sized grenades.

Perhaps the most remarkable of all mortars was the Second World War Soviet 37-mm Spade mortar, which was exactly what the name implies: it was a mortar inside a spade. When required, the handle was detached from the spade blade, which became the base plate; it could then fire a 1½-lb bomb 328 yards. If not required as a mortar, it could be used as a spade. The disadvantage was that it weighed 5.3 lb and, as the weight which a soldier can carry is carefully calculated and distributed, this particular piece of firepower, though ingenious, was not considered sufficiently valuable to form so large a share of the soldier's burden.

Although many of the ideas for making guns more powerful, smaller and more adaptable were rejected by War Departments, which appeared to believe in the principle of 'better the devil known', the railway gun, first used effectively in the First World War, had an instant appeal. Formerly, the amount of recoil would have made the whole idea of such a gun ludicrous, but now that control of recoil had been mastered there seemed no reason why a powerful gun should not be able to move freely along railway lines, fire at a target, move on to another position, fire again and then retreat out of range. As armies found out when they put the idea into practice, there was in fact a reason why a railway gun might not perform as well as had been predicted: its field of fire was too restricted. When a railway gun arrived at its appointed destination, a suitable system of jacks and beams was put around the truck on which it was mounted; these took the strain off the gun platform. However, there was very little freedom to swing the barrel of the gun, which had to remain facing the general direction in which it had been travelling. For the smaller guns this problem could be overcome by halting the truck on suitably positioned turntables, but a bigger gun required more space for its recoil than a turntable could provide. Thus the smaller railway guns became fully traversible when turntables were available, but the larger ones needed a greater length of line than could be provided on a turntable. However, the fact that the larger guns had to stay on the main track eventually enabled the Germans to turn a necessity into a virtue. When possible they halted their guns close to a tunnel. After firing they were drawn back inside; there they stayed, concealed from enemy aerial observers and other curious eyes, until needed again.

But the main use of railway guns was not on the battlefield at all. During the First World War both Britain and Germany were well aware that at any moment their homeland might be invaded by a force escorted by the other side's powerful navy. The plan might be either for a full-scale invasion or for a raid on a major port. Britain had plans for approaching Germany via Texel, Arnhem and Essen; via Wilhelmshaven; and via Kiel. Britain felt vulnerable at Sunderland and Hartlepool, at Whitby, at Hull, at Cromer and at Harwich. Naval forces were earmarked for the defence of these points, but one could never be sure what would happen if an enemy force came in through fog or bad weather after a successful decoy operation had lured the assigned defenders elsewhere. In those circumstances a few railway guns, rushed to the spot to supplement the existing coastal batteries, would be of untold value. In the event no invasions took place, but there were indications of what might happen when both Hartlepool and Scarborough were shelled by German ships, which killed several hundred civilians. In the Second World War thoughts of the Germans mounting a seaborne threat were discounted after 1941, although when two of their capital ships slipped through the Channel unobserved the Higher Command had reason 'to suck its teeth', as the Navy like to put it.

All the inventive ability, managerial skill and unceasing labour which went into the manufacture of armaments had, of course, the single objective of destroying human lives. Those working in ordnance factories did not perhaps see their jobs in such straightforward terms: they were 'helping our lads at the front', 'battling for Britain' (or Germany), or 'feeding the guns'. Many of the instruments of death which they manufactured were so precise and well-proportioned that they were almost objects of beauty, and even looked harmless. But for those who faced them, how different. In *Ypres 1917* Norman Gladden wrote:

> The enemy barrage now seemed to lift and fall with full force on the wood. Out in the front a red light shot low, fell over the ridge and burnt itself out on the ground, suffusing the waste of shell-

holes with a pale crimson glow. Other lights followed and our guns redoubled in intensity as they replied to the SOS call. The enemy was attacking all along the line.

The wood was now in terrible upheaval. From above, spiteful shrapnel shells burst downwards, spattering the branches with bullets, while high explosives burst below, causing here and there a tree or large branch to fall with a crash. The shells were falling so quickly that the aisles behind were filled with dense fumes. I clutched my rifle expecting every minute to be my last. On all sides I could hear cries and groans and could only marvel that anyone survived.

Ernst Junger, a nineteen-year-old lieutenant in the 73rd Hanovarian Fusilier Regiment, was less than a mile from Gladden. In his diary he wrote:

In the early hours of the afternoon we were heavily shelled with shells up to the heaviest calibres. Between six and eight o'clock one explosion overlapped another, and the building often trembled at the sickening thump of a dud and threatened to collapse. When the fire ebbed later on, I went cautiously over a hill that was covered in a close and whirring mesh of shrapnel to the Kolumbesei dressing-station and asked for news of my brother. The doctor, who was examining the terribly mangled legs of a dying man, told me to my joy that he had been sent back in a fairly promising state.

I could not see a great deal, as the surroundings were veiled in a thick haze. The artillery fire was increasing from moment to moment, and soon reached that pitch of intensity which the nerves, incapable of further shock, accept with almost happy indifference. Showers of earth clattered incessantly on the roof, and the house itself was hit twice. Incendiary shells threw up heavy milk-white clouds, out of which fiery drops rained on the ground. A piece of this burning stuff came smack on a stone at my feet, and went on burning for a good minute. We were told later that men hit with it had rolled on the ground without being able to put it out. Shells with delay-action fuses burrowed into the ground, rumbling and pushing out flat discs of soil. Swathes of gas and mist crept over the battlefield, hugging the ground.

> Immediately in front we heard rifle and machine-gun fire, a sign that the enemy must have advanced already.

In France and Belgium, the weather seemed to those exposed to it to contain more periods of rain than they had previously believed possible outside a tropical monsoon area. This made life unpleasant enough for the infantry, but at least they only had themselves to move around in the slimy liquid mud which resulted from rain and incessant shellfire. The gunners had the additional task of moving huge pieces of machinery which were difficult enough to handle on dry ground: on the Western Front mules, guns, limbers, ammunition wagons and men all sank into the glue-like mess underfoot.

Mud was all too familiar, but a constant source of surprise was the string of new devilries produced by the enemy. To a gunner one shell was like another, though there might be a slight difference in name or shape. For the infantryman of either side it was a different matter:

> In May the enemy began to use a new fuse with great effect. Its instantaneous action caused the shell to burst directly it touched the ground, sending a shower of splinters in a forward semi-circle to a distance of several hundred yards. It was, of course, quite useless against dug-outs and shelters, but was extremely effective against troops in the open. In addition, shells fitted with the instantaneous fuse had a greater moral effect on account of the deafening crash of their burst. The arrival of a salvo can only be compared with the rapid rending of enormous tea trays by some angry giant. They produced no shell-hole. A lightly-scraped cavity in the ground was the only indication they left behind.

A sense of humour is not normally considered as a factor in firepower but, as every soldier knows, the ability of a soldier to turn a frightening situation into a joke is an invaluable factor in morale – and morale is firepower:

> Gunner Pope, the cook, a burly ruffian, whose face and hands and clothes are uniform in colour with his own stewpots, is engaged in cooking the mid-day meal behind the shelter of corrugated iron, known by courtesy as the cook-house. Two satellites, only one degree cleaner than the Master himself, are

busy peeling potatoes. Suddenly a large shell, fortunately a 'dud', falls in the middle of the little group, and cook, potato-peelers, dixies and stewpots are sent flying. Gunner Pope extricates himself from under the heap of corrugated iron sheets. Rising carefully to his feet, he ruefully regards the wreckage, 'That's the third time this week the spuds have been ruined by those blasted Allymonds.'

An officer writing in 1916 and sheltering behind the anonymous signature of F.O.O. recorded:

> Observation posts have each their own legend, which clings to them through successive tenancies. We shared one once with a very youthful officer whose nervousness was only exceeded by his ignorance. One fatal afternoon he had to observe a series [gunnery practice]. The first round was fired and the young man, suddenly discovering that observation of fire is one of the most difficult things in the world, hesitated on his report.
>
> The battery commander came on to the telephone. 'Ask the observing officer to report where that round fell.'
>
> Orderly: 'Mr Jones reports that it was a very good shot, sir.'
>
> 'Tell Mr Jones I don't want criticism of my shooting, I want to know where the rounds fall. No. 2 is just firing.'
>
> 'Mr Jones reports the last round fell about an inch from the target.'
>
> 'Then I can assume it is a hit,' replied the battery commander sardonically.
>
> 'No, I don't think so, sir, Mr Jones says he means an inch on the map not an inch on the ground.'

These were early days. No doubt if Mr Jones survived the subsequent interview with his battery commander, and a few more months of action, he became a very experienced observation officer. If so, he probably did not survive the war. Casualties among forward observation officers were said to be higher than for subalterns in any other arm of the service, doubtless because they were so often in exposed positions. In the Second World War this unenviable position and statistic was taken over by Brigade Signals officers who, if not always observed, were easily located by the enemy.

7

Tank and Anti-tank

The tank or, as it is more correctly designated, the armoured fighting vehicle, is commonly thought of as a weapon which belongs entirely to the twentieth century. However, along with many so-called modern weapons it has a long ancestry.

In pre-Christian times the function of the modern tank on the battlefield was performed by war chariots, first recorded as being used in what is now Iraq in 3000 BC. Soon they were included in the armies of countries as far apart as China and Egypt, as may be observed on many ancient sculptures and friezes. They were drawn by semi-wild horses or asses, were often armour-plated and were principally used for making sudden forays. They did not, as is often assumed, have spikes or knives protruding from the wheel-hubs, for that would have made them unmanageable. The Ancient Britons who fought against Julius Caesar in 55 BC had sturdy, metal-protected, ornate chariots, but the feature which impressed the Romans most was the skill and agility of the charioteers. The chariots hurtled on to the enemy lines, sweeping away opponents by their weight and speed; the charioteers slashed and hacked at anyone within reach, often leaping in and out of the vehicle when they thought the moment appropriate. Commanders of light modern tanks would have found much that was both familiar and admirable in the tactics of Celtic charioteers.

The Italians seem to have been first in the field with a design for a wind-driven vehicle – a sort of armoured land yacht – in 1472. In 1599 the Netherlanders were experimenting with the same idea. Thirty-five years later an Englishman had sufficient confidence in his own design for a 'landship' that he patented it. All these craft were hopelessly impracticable for the battlefield because of the unreliability of windpower, but the invention of the steam engine made the outlook for landcraft more promising. In 1770 the French Army produced a steam vehicle which would travel at $2\frac{1}{2}$ mph, but there were very few ground surfaces on which it could travel at that speed – or even at all. A more important, though less spectacular, invention of those days was a form of caterpillar track. The original version consisted of wooden slats which were laid down in front of the vehicle and picked up when it had passed over them. Thirty years later an improved application of this idea incorporated the slats into chains on the vehicle; however, the mechanical problems were too much for the engineers of the day. Another eighty years passed before further experiments took the embryonic tank further. These partly overcame the problem of steering a tracked vehicle and also enabled it to surmount obstacles by having its tracks sloping upwards at the front.

By the end of the nineteenth century the petrol engine had been invented and made possible that close companion of the tank, the armoured car. The French had a model built by Charron et Cie in 1902; the Germans had one built by Daimler a year later. Soon armoured cars were included in the armies of many nations, and the Italians used them to control their colonies in North Africa in 1913. Meanwhile the idea of the tank – an armoured, tracked fighting vehicle, had remained in abeyance. An English inventor called Cowan had patented an armoured vehicle based on the steam tractor: this, however, had wheels rather than tracks. A similar vehicle, produced by the British engineering company of John Fowler, appeared in time for the Boer War but was used only as a tractor and not as a fighting vehicle. In 1904 the designers at Charron had expanded their armoured car into a vehicle with a turret. Daimler produced a similar vehicle in Austria in the same year.

It seems extraordinary today that the potential of the armoured fighting vehicle was not immediately recognized and examples used in the early years of this century. In fact, the neglect was entirely predictable. The tank may be master of the modern battlefield, but until the First World War the horse was regarded as the only respectable means of mobility in the face of the enemy. Cavalry was thought to have ruled the battlefield since the Battle of Hastings in 1066, and perhaps earlier. When in the Edwardian era senior officers spoke of war they thought of Prince Rupert, of Marlborough, of the

Greys charging at Waterloo, of the Light Brigade in the Crimea, and more recently of French at Klip's Drift. Occasions when cavalry had been rather less successful, in the face of pikes or arrows, were tactfully glossed over or ignored completely. In 1903 the French turned down a proposal for a tracked vehicle which would carry at least one large gun, and in 1908 the British took the same decision. A more advanced design was rejected by the German General Staff and their allies in 1910. The British War Office received detailed plans in 1912, looked at them and filed them, so that they were not seen again until many years later. It was rumoured that someone had written across the front page, 'This man is mad.' The neglect of these most valuable plans came to light in 1919, when the inventor applied to the Royal Commission on Awards to Inventors. Unfortunately, the fact that the tank had developed along the lines he had predicted did not, they decided, justify an award to L. E. de Mole, whose plans would have saved years of experiment if they had been adopted.

By 1915 it had become clear even to the most blinkered cavalrymen that some form of armoured vehicle was necessary if machine guns were to be mastered, trenches crossed and wire flattened, none of which feats could be accomplished by horses. Unfortunately the first designs for a landship were much too large: one was said to weigh over 1000 tons.

Once the idea of a landship had won grudging acceptance it was necessary to commission someone to make the necessary experiments and construct the ultimate vehicle. The only organization with experience in armoured car construction was, surprisingly enough, the Royal Naval Air Service, which used them to protect those of their squadrons which were based on Dunkirk. The First Lord of the Admiralty, under whose aegis the RNAS flourished, was Winston Churchill, who was keenly interested in the idea of the tank as a means of breaking the deadlock of trench warfare. The first vehicle remotely resembling a practical tank appeared in July 1915 and consisted of an armoured car body mounted on a Killen-Strait tractor; it was built at William Foster & Co.'s factory at Lincoln. Though driven by a 105-hp Daimler engine it could not manage to cross a ditch more than 4 feet wide. Nevertheless it proved that an armoured landship was possible. In order to keep the project completely secret the staff at Foster's, who were aware that something new and important was going on, were told that anyone curious to know how they were occupying their time must be told they were manufacturing water tanks. Thus the 'tank' was born, though this first one also acquired the affectionate name of Little Willie.

Little Willie, which had been specially manufactured to solve the Army's problems, had in fact been built without any army specifications being submitted until they were too late to be incorporated. An interesting situation had developed: the Director of Naval Construction was supervising the manufacture of tanks for the Army, which had not requested them and was doing nothing to help in their construction. The man supervising the project was a young naval officer who was neither a regular nor an engineer: in peacetime he was a banker; nevertheless the evolution of a practical tank owes much to Lieutenant A. G. Stern, RNVR. In September a second tank put in its appearance. This was Big Willie, built in a lozenge shape to enable it to fall into ditches and then climb out of them. Unfortunately the fact that the tracks went over the top of the vehicle meant that, at this stage of development, the guns could not be mounted on top. They were therefore put at the sides in what the Navy terms 'sponsons' – projections from the sides of warships, enabling guns to be trained fore and aft. Tanks which carried a 6-pounder gun were known as 'male' tanks, and those with machine guns only were 'female'.

Big Willie completed trials successfully by December 1915; the prototype, known as Mother, is now in the tank museum at Bovington in Dorset. Having proved that it could cross wide trenches, Big Willie received a nod of approval from the British Army and 100 were ordered two months later. Meanwhile the French, who had been engaged in similar experiments, produced their version of a tank at the same time. It was less sophisticated than Mother, but the French Army had sufficient confidence to order 400. The progress of these two allies with their tanks makes an interesting contrast. The British tanks were brought into action for the later stages of the Battle of the Somme, when forty-nine were used on 15 September 1916. Their success was limited because they were badly employed tactically and much of the element of surprise was wasted. They did much better at the Battle of Cambrai on 20 November 1917, where 474 tanks were used and broke through the German lines in spectacular fashion. Coming immediately after the long, grinding Battle of Passchendaele, when thousands of men had toiled through mud to gain 3 miles, the Battle of Cambrai seemed to herald the dawn of a new era. For a similar gain, fewer than a hundred men had been lost. The first thrust, made by 381 tanks followed by a small number of infantry, had taken the Germans completely by surprise. Unfortunately for the British their tanks were too slow, having a top speed of 4 mph, and their 30-mile range too limited to exploit the breakthrough. The Germans were lucky in that a reserve infantry division had just arrived back from the Russian Front and was available for the counter-attack. In fact the Germans had much more luck than they deserved at Cambrai. Their General Staff had been perturbed when British tanks appeared on the Somme a year earlier, but had soon recovered its confidence. The new weapon, it was decided, did not constitute any real threat. It was too slow, too clumsy and too prone to breakdown. Even when a prisoner captured from an Irish regiment in November 1917 had given away the secret of the impending tank battle at Cambrai, the German Staff was so confident in its own judgement that he was not believed, even though his information was quite detailed. The final point which convinced the Germans at Cambrai that they were in no danger of attack was the fact that there was no preliminary bombardment. Everybody knew that if there was going to be an attack ample warning would be given by a lengthy bombardment; conversely when there was no bombardment there could be no attack.

The tanks, which had been rehearsed in the necessary procedures, went in in groups of three. The commanders knew that three wide trenches had to be crossed in order to drive a wedge into the German lines, so they resorted to a device which had been successfully employed by Marlborough over two hundred years earlier. The tanks carried large bundles of brushwood, known as fascines. As the leading tanks reached the first trench they slipped off their fascines, thus filling up the ditch. The second line of tanks then passed the first and gave the second line of trenches the same treatment. When the third trench had been passed, the Germans had only skeletal defences between the advancing British and the open country behind. There was a wide gap between Masnières and Crevecoeur. Tanks had been lost, but there were still others to take their places if they were brought forward in time. In addition there were two cavalry divisions which for the whole war had been waiting for a moment like this. But the British attack had slackened, bemused by its success. Meanwhile the Germans were rushing up a further five divisions, with six more under orders to follow if necessary. The German Air Force was ordered to support the counter-attack with all-out dive-bombing, and did so. It was a foretaste of the 1940 Blitzkrieg. By the end of the month all the British gains had been lost and only with great difficulty had the German counter-attack been halted.

It was a bitter blow for Haig who, as Commander-in-Chief of the British forces, had always looked forward to the day when his cavalry could pour through a gap in the German lines and turn victory into rout. He had never had great faith in tanks, but now they had lost after an initial victory he took all the blame himself. The thought of the war-winning potential of a break-out by the cavalry had made him batter away for months to gain the Passchendaele Ridge, which he had finally reached too late in the year to be of value; now that an even better opportunity had been presented to him unexpectedly, it had proved an opportunity lost. For the second time the advantage had been wasted. In the subsequent enquiry over how this had happened, blame was placed on the inadequate training of officers and men. But, as armoured units would discover later, there is a definite limit to what men and machines can achieve, even in victory.

The Germans probably learned far more than the British from the tank battle at Cambrai. On the Somme they had been surprised but not over-impressed by the new weapon; the General Staff pronounced that it was bad for morale, both for the tank crews who were penned up in their iron boxes, and for the infantry who could easily come to place too much reliance on the new weapon. At Cambrai the Germans were presented with an unmistakable example of how the new weapon should not be used – attacking along too wide a front, having insufficient reserves for the main thrust, and missing the opportunity to exploit gains. With this they had discovered – from necessity – how low-flying aircraft can be used to support ground troops when the battle is mobile. As soon as the treaties which ended the First World War had been signed the Germans were practising in secrecy, and cogitating on how those lessons could be translated into the tactics which would enable them to invade France in the Second World War.

What Cambrai taught France and Britain was that lighter, faster tanks were needed. In the spring of 1918 the first batch of these Medium A or Whippet tanks went into action. They had a top speed of 8 mph and a range of 80 miles. On 26 March seven Whippets broke up a German attack, killing nearly four hundred men for the loss of twelve. Tanks had clearly justified themselves. Somewhat surprisingly, although the British had done nearly all the pioneering and the fighting, the most popular tank was a French Renault. Though light and slow, it was manoeuvrable and reliable; it was also a subordinate weapon. The Renault's job was to work with the infantry, which it was felt was the most important arm on the battlefield. Artillery could shell targets, aircraft bomb them, engineers blow them up, but if wars were to be won the infantry had to stand firmly on the contested ground. Tank tactics in the French, British and US armies would henceforth be designed to assist the infantry, not to conduct a campaign on their own. Meanwhile, with the lessons of Cambrai firmly in mind, and with the writings of British tank experts to confirm their view, the Germans were setting to work to build a tank force which would slice a dozen holes in the enemy lines and then flood through them in 'an expanding torrent', a phrase originated by Major General J. F. C. Fuller but used by many other writers.

Although the British had been the pioneers of tank design, by the end of the First World War they had only built 2636 in contrast to the French output of 3870. The majority of French tanks were the Renault FT models, the designs of which were copied by the Italians and the Americans. Although the Germans had found out so much about tank warfare from the Battle of Cambrai, they never entered into serious tank production before the Armistice: their total wartime output was less than twenty, and when they analysed the reasons for their eventual defeat lack of tank production was seen to be only one of them.

Although restricted from having tanks by the provisions of the Treaty of Versailles, the German Army was able to experiment with the tactical possibilities of the new weapon by using models manufactured from cardboard and wood. Most of the senior officers still serving in the German Army nursed considerable doubts about the value of the tank as a weapon, but General Hans von Seeckt was not one of them. Seeckt was a keen student of military history and believed – as Schlieffen had believed – that an army capable of surprising its enemy by swift encircling movements was likely to be victorious. When Commander-in-Chief of the German Army between 1920 and 1926 he was able to give much secret encouragement to the tank pioneers. In this period the Swedes sought favour with what they felt was soon to be a resurgent Germany, and built tanks to German specifications in Swedish factories. The Russians, pursuing a policy which eventually proved to be far from enlightened self-interest, provided training facilities both for the German Army and the illegal German Air Force near Moscow (see Chapter 11). Meanwhile, a young captain named Heinz Guderian, who had been given a staff appointment in the transport division of the German Army, wrote an article on the possible use of armoured vehicles in a future war. In 1923 Walter von Brauchitsch, later destined to be Commander-in-Chief, tried out some of Guderian's ideas on manoeuvres. These ideas were, however, by no means entirely original: Guderian had drawn many of them from the writings of the two British military theorists B. H. Liddell-Hart and J. F. C. Fuller. Unfortunately for Britain, Liddell-Hart and Fuller were considered too venturesome and unorthodox by those who determined policy in the War Office. When exercises with what was described as Experimental Mechanized Forces were held in 1927 and 1928, the report on them was not taken very seriously by the senior military pundits. It was, however, studied very carefully and found very useful in the German Army. Nobody outside Germany knew that the Wehrmacht had already received two real tanks, manufactured for them in Sweden, one weighing 9 tons, the other 20. Others would follow. When, in 1931, Guderian was made Chief of Staff to the German Commander-in-Chief, Oswald Lutz, the foundations of the German Panzer armies had been finally laid.

The word *Panzer*, which the outside world soon attached to the whole concept of Wehrmacht mobile power, simply means 'armour' in German. It is used to make compound words such as *Panzergruppe* (command unit), *Panzerjäger* (anti-tank troops) and *Panzerkampfwagen* (armoured fighting vehicle). In the German forces the term Pzkw was used – an abbreviation of *Panzerkampfwagen*. Even though the German armoured vehicles ran into considerable trouble as the Second World War progressed, and came off much the worse after meeting the anti-tank weapons of other armies, they never lost their mystique of invincibility.

In the years immediately before the outbreak of the Second World War army commanders encountered a new range of problems. The tank or AFV (armoured fighting vehicle), as it was usually known, had made its début as a blunt instrument for demolishing sophisticated enemy defences. Its potential was strictly limited by its low speed, its great weight, and its lack of manoeuvrability and offensive firepower. Over the years much progress had been made in reducing all these handicaps, but as armies became more experienced in the ways of the AFV it became obvious that its valuable qualities must all be assessed in relation to each other and to their collective value on the battlefield. If the offensive weaponry carried by the AFV was to be increased, how far could this be achieved without adding too much weight? Inevitably there were strong clashes of opinion. Some commanders believed that speed was all-important, and as the AFV could never carry a sufficient thickness of armour to make it invulnerable to a direct hit then armour should be reduced to the point at which it gave no more than adequate protection against small-calibre fire. The weight thus saved would allow the vehicle to move more rapidly and flexibly, two features which would enable it to avoid hostile fire and to play a more useful part in battle tactics. Furthermore, if the AFV carried a larger gun it could probably outrange the enemy anti-tank weapons and destroy them before the tank lumbered into their sights. Unlikely, said the proponents of thicker armour: a gun light enough to be carried on an AFV would be lucky to find an anti-tank or field gun which it could outrange.

The debate continues to the present day, and will no doubt last as long as AFVs are deployed on the battlefield. Bigger and better guns fire off their large ammunition faster, but before they do so it has to be stored inside the tank, occupying space contended for by fuel and, of course, those essential human beings who are known as 'tank crews'. Until robot tanks are universal – if ever they are – the tiresome qualities of the human being have to be taken into account. He needs space, has to be able to breathe, is liable to become 'seasick' and therefore useless if the ride is too bumpy, needs periodically feeding and watering, and must not become too hot, too cold, too tired, too reckless or too unenterprising. Modern AFVs have crews of four, crammed into a space easily filled by two.

In spite of its appearance of complete invulnerability, an AFV is accident-prone. It can be immobilized quite independently of enemy action: it can fall off cliffs, sink into mud, fail to climb beaches, break down through overheating or some other cause, damage a track . . . even its predecessor in warfare, the horse, would be hard pressed to produce so many reasons for not being able to perform adequately on the battlefield.

But even this list does not exhaust the drawbacks to the use of

LEFT: A siege in the Middle Ages. Both attackers and defenders are using siege engines (trebuchets) powered by counterpoises to project rocks and incendiary material. The longbowman in the right foreground has been transfixed by an arrow from a ballista. A crossbow bolt is stuck in the ground next to a crossbow.

BELOW: English longbowmen confronting French crossbowmen (on left) on a 14th-century battlefield. The crossbow was more powerful but could only discharge approximately two bolts each minute; the longbow could send off twelve in the same time.

BOVE: Fortification during the Crimean War, 1854–6, ter the battle. This strongpoint was held by the ussians but was captured by the British after bitter ghting.

GHT: Attacking a fortified town in 17th-century urope. The wickerwork cylinders (gabions) were filled ith earth and could be moved forward or back. The edieval walls were screened by a defensive outwork ned with artillery.

RIGHT: A Fokker E1 in 1915. This Dutch design had been rejected by the British but was quickly taken by the Germans who used it to great effect.

BELOW: Major Bill Barker, British fighter ace, with his Sopwith fighter.

BELOW: World War I transport driver with his horses, all three wearing gas masks.

OPPOSITE ABOVE: The Battle of the Somme, August 1916. British howitzers in action in the Fricourt area.

OPPOSITE BELOW: Germany, forbidden to have tanks by the Treaty of Versailles, 1919, improvised models with wood and cloth and used them to practise tactics.

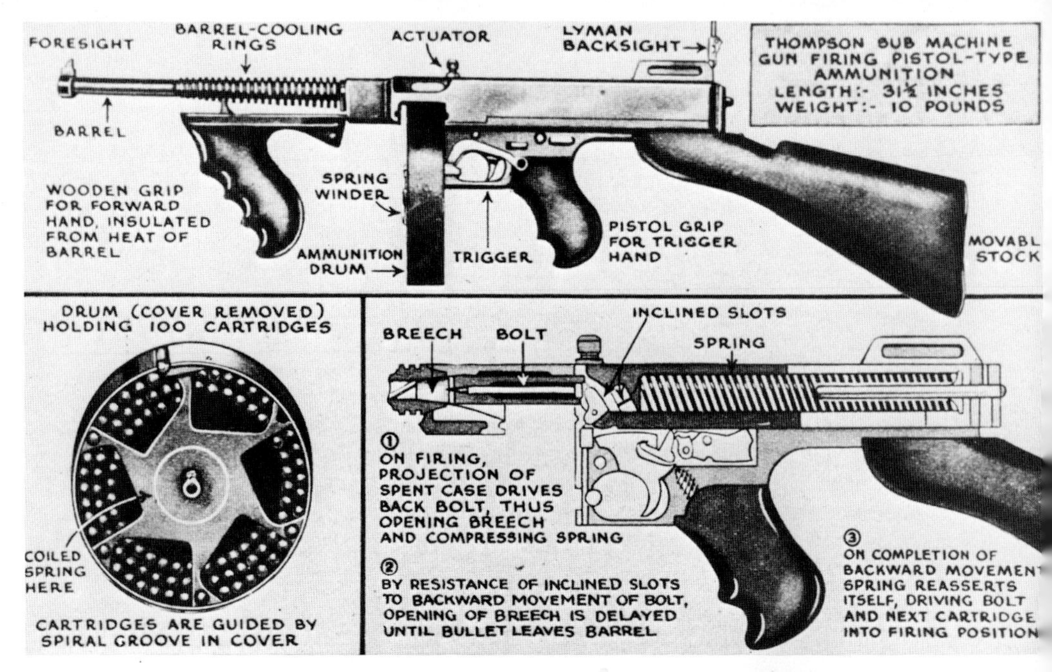

ABOVE: The 1940 'Tommy gun'. The Thompson sub-machine gun, shown in section above, was devastating in close-quarter fighting but the ammunition it fired so rapidly was heavy and bulky and therefore a handicap logistically.

BELOW: British soldiers loading a 5.5 gun to engage an enemy target which has been located by sound waves recorded on microphones, and flash spotters from aircraft.

ABOVE: The Heckler and Koch 33E, a popular automatic rifle. Calibre 5.56 mm, 25-round box magazine, semi-automatic and sustained fire.

BELOW: British soldiers of the 2nd Cheshires with .303 Vickers machine-gun at Andrieu, Normandy, in 1944. The town still contains Germans and the two soldiers are watching closely for any sight or sound of enemy movement.

RIGHT: Sherman tanks in action against the Germans at the battle of Alamein in October 1942. Shermans were American tanks manufactured from a British design.

BELOW: A German PZKW IV knocked out by Russian guns in the battle of Kursk, 1943.

LEFT: A British 'Scorpion' mine-clearing tank equipped with flail chains which exploded mines harmlessly instead of under the tank. A brilliant British invention which was invaluable in Normandy and in the Western Desert.

BELOW: 'RPG7' Soviet anti-tank rocket and launcher, used world-wide.

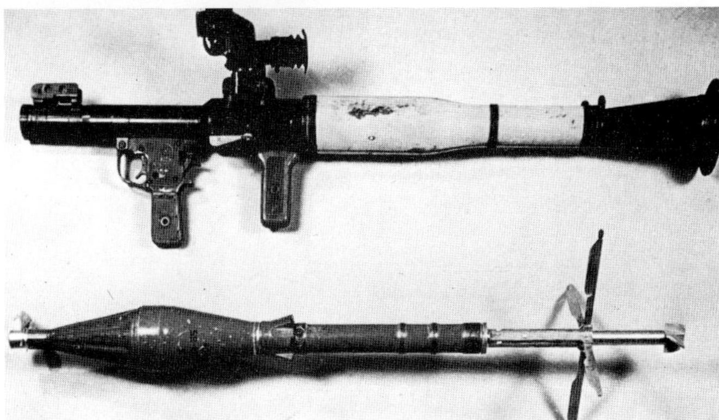

ABOVE: A German 'V2' rocket. A thousand of these were launched onto London; three thousand elsewhere in Britain. The warhead carried a ton of high explosive. It travelled faster than the sound it made and therefore arrived without warning.

ABOVE: A British 'Seacat' surface-to-air missile being launched.

RIGHT: The Exocet AM39, an air-launched anti-ship missile of French origin in company with a French Mirage F1 fighter.

ABOVE: A Soviet T-80 Main Battle Tank.

LEFT: M60 A1. Main Battle Tank of the US Army.

LEFT: USS *Iowa* firing at a shore target during the Korean War, 1952.

BELOW: USS *New Jersey* first saw action in World War II in the Pacific, served in the Korean and Vietnam wars, then was modernised and rejoined the fleet in 1984. Note 16-inch guns.

BELOW: USS *Nimitz*, nuclear-powered aircraft carrier accompanied by nuclear-powered guided missile cruisers, USS *California* and USS *South Carolina*.

ABOVE: Seven ALARM missiles carried on a British Tornado GRMK1.

LEFT: Soviet MiG-31 'Foxhound' long-range interceptor fighter.

BELOW: Grumman E-2C Hawkeye electronic surveillance aircraft.

ABOVE: Rockwell B1 strategic bomber. Range over 700 miles at 600 mph.

RIGHT: Dassault-Breguet Mirage F1 (attack/strike aircraft) launching a salvo of rockets from its four 18-round launchers.

LEFT: Four Jaguars from No. 6 Squadron RAF Coltishall on a low-flying exercise.

TOP: 'HIND' attack helicopter of Soviet Air Force.

ABOVE: HARM (High-speed Anti-radiation Missile) locks on to its target by homing in on its radiation. Here mounted on USAF F-4G (Wild Weasel) aircraft.

US Navy 'Tomahawk' cruise missile launched from flat-based truck.

AFVs. To train an infantryman needs little except strong boots and adequate supplies of ammunition; to train a tank crew requires large quantities of fuel. On the battlefield an AFV will display a voracious appetite for fuel and ammunition and, while moving to the point at which it will be most useful, it will help to destroy roads which are needed for other tanks and other users. Its presence will add enormously to the congestion on the battlefield and thus will provide excellent targets for enemy aircraft and artillery. Its final weakness is that it cannot operate without infantry, for the latter are needed to consolidate and defend ground which has been over-run.

Nevertheless, despite all its drawbacks and vulnerabilities the AFV is a battle-winning factor. With the conviction that this was likely to be so, in the 1930s the German Army set about building a Panzer force which would eventually slice through opposing armies in Poland, Belgium, France, Yugoslavia, Greece, North Africa and the Soviet Union. Realizing that tanks needed their infantry back-up, who could not be expected to march at 25 mph, let alone cover the 100 miles that a tank would range in the Second World War, the German Army decided that its 'foot sloggers' were now to be transported in half-tracked personnel carriers (*Schützenpanzerwagen*). The AFV moving into action would in some respects be like a hen with its brood of chicks. There would be infantry, ammunition and fuel supply lines, and mechanics who would keep one's own tanks moving but also be on the alert for any serviceable enemy AFVs which had been abandoned on the battlefield but could be salvaged and used against their former owners.

Supplying spares for tanks was no simple task: by 1939 an AFV had ten thousand separate parts, and by the end of the war this figure would have increased threefold. And, as AFVs became more and more complicated, various problems in connection with them reared their heads far distant from the battlefield. They were problems of design and production. In order not to present too good a target, the outline of the tank needed to be as squat as was consistent with efficiency. But AFVs were obviously going to be hit at times, and it was therefore important that their armour should minimize the effect of enemy shells by deflecting them. (This was one of the features of the Russian T34 tank which would cause the Panzers considerable trouble later.) The problem of production was also closely linked to success on the battlefield. As with an aircraft, the most critical decision was when to put a design into production. Could it be assumed, before assembly lines were laid down, workers trained and components assembled, that this design would remain serviceable for several years? If a design had some inherent flaw – such as a blind spot, or propensity for the gearbox to overheat – this would not be discovered until it was tested on the battlefield. Trials under peaceful conditions were valuable, but never produced the complete answer. Training grounds in Britain or Germany could hardly reproduce the rigours of the Russian winter or the African desert, but something had to be produced which would cope reasonably well with either. Yet one mistake over the timing of starting production could lead to what might become an immediate coffin for hundreds of air or tank crews.

Whatever other problems tank design and manufacture posed, the most vital was to give it bite: in other words, a powerful gun. When Hitler took over the Skoda factory in Czechoslovakia in spring 1939 he acquired 469 tanks, each of which carried a 37-mm gun and two machine guns. The possession of these tanks, and the facilities to make as many more as were required, virtually ensured the German victories in France a year later. Later German tanks carried 50-mm, 75-mm and 88-mm guns. The Tiger I had an 88-mm gun, and the Tiger II mounted one of 128 mm. Even these powerful monsters were outclassed by the self-propelled guns (*Sturmgeschütz*), which were carried on a tank chassis but not mounted in a turret and therefore could not swing in a 360° arc, but only in one of 20°. For its intended purpose the smaller angle was satisfactory. The gun was aimed by the driver and the gunner zeroed in with his telescopic sight.

In the course of time and experience many forms of tank were designed and used. There were anti-aircraft tanks which carried guns suitable for defending armoured columns, there were tanks which carried flame throwers, there were tanks carrying radios which were exclusively designed for communications work, and there were command tanks occupied by those in senior positions – but not necessarily much more comfortable for that. There were mine-clearing and rubble-clearing tanks with great shovels in front of them (as well as additional armour), weapon-carrying tanks and petard tanks (named after the explosive charge once carried by engineers, as in the expression 'hoist with his own petard'), which carried a howitzer for short-range demolition. The British displayed a genius for inventing unusual tanks, such as the flail carrier which exploded mines by means of whirling chains, and the amphibious vehicles known as DD tanks, which could propel themselves through water. But the value of the AFV, as with every other weapon, depended on the skill and daring with which it was used. Rommel displayed great dash and ingenuity when he manoeuvred his tank columns in the Western Desert, hooking them like a steel claw around the opposing flank or sometimes driving a wedge through the middle of a straggling formation. But an AFV had to be able to look after itself. In July 1943 the German Army launched an offensive against the Russian salient around Kursk, some 250 miles north-east of Kiev. They had a tank force of nearly 3000, which included the new Tigers and Panthers of which much had been expected. In nine days the Germans had lost 2000 tanks, 1300 aircraft and 70,000 men killed. The new tanks had proved a great disappointment: some of them had been knocked out by Russian artillery, others captured by infantry. There were, of course, ancillary reasons for this German disaster in the greatest tank battle of the Second World War. A violent storm had created a swamp out of land which had been dry the day before, and the combination of intricate minefields and determined Russian resistance had brought the German tanks to a standstill. Ammunition, too, had run short, but that might have been expected.

In view of the fact that in the Second World War the Soviet Union and Germany made more use of tanks than other nations, it is interesting to remember that both derived most of their knowledge from the British Vickers Medium Tank of 1923 (16 mph, range 150 miles, 45-mm gun, 6.5-mm armour) and the American Christie (30 mph, 9 tons, and armour varying according to model).

As early as 1932 the Russians had built an advanced model tank, which carried a 76-mm gun. They were the first nation to realize that a tank requires great hitting power and, while the Germans still had faith in the 37-mm and 50-mm for the Pzkw III, the Russians were equipping their T34s with 76-mm guns. Britain lagged far behind: 80 out of the 1100 British tanks were armed with a 40-mm gun, and the remainder had to rely on machine guns. The Italians were in no better a situation, and the United States was even worse off. The French had 2500 tanks of which 150 were Char Bs carrying a 75-mm gun. In view of the fact that the Soviet Union was generally thought to have been caught unprepared for the Second World War, it is interesting to recall that at the outbreak, and thus nearly two years before she was invaded by Germany, Russia had 20,000 tanks, a total exceeding that of all the other tanks in the world: the German total of 3100 tanks in 1939 was pitiful in comparison. Nevertheless it is not, of course, merely numbers which count in battle, but the way in which they are handled. The Russians quickly learned the most effective way to handle tanks in battle, and they matched their enterprise in the field with their design: the KV1, a 46-tonner, was far superior to the German Pzkw IV. When the Germans produced the Tiger, armed with an 88-mm gun, the Russians equalled it by arming their T34 with an 85-mm gun, and then produced the Josef Stalin tank which carried a 122-mm gun. The Germans never gave up the contest of 'my tank is bigger and better than yours', but their 'ultimate weapon', the Maus, armed with a 128-mm gun and with 350 mm of armour plate to protect it in front, could manage a top speed of only 12 mph and therefore had very limited use.

British tanks had a far from happy career in the early days of the Second World War, even though there were notable thinkers among the proponents of armoured units. Captain G. Le Q. Martel had written a prophetic paper on tank warfare in 1916, J. F. C. Fuller had taken up the cause, and Percy Hobart had produced brilliant, far-ranging ideas, but the British Army still saw the tank as a means of infantry support. These original roles had now been reversed, but the powers that ran the War Office were not prepared to see it. When the war began, much was hoped for from the Matildas, Crusaders, Cromwells and Covenanters, but the story soon became a dismal recital of breakdowns, under-gunning, low speed, poor tactical use and inadequate armour. To add insult to injury, there were not even enough of them. The last problem was resolved dramatically, as shown in the table of comparative production figures for Germany, Britain, the USA and the USSR.

As the war progressed there was much interchange of weaponry among the Allies. The USA sent the USSR vast quantities of vehicles, oil and raw materials; Britain sent aircraft to the USSR, and Britain and Canada together supplied 79 per cent of the shipping used when the Allies landed in France in June 1944.

The way in which the balance of firepower is affected by what might appear a minor change was shown by the tank battles in the Western Desert during 1942. The British tanks initially carried a 2-lb gun, which the War Office believed, on the basis of tests and exercises, was

German and Allied tank production figures (approximate), 1940–4

	Germany	*Britain*	*USA*	*USSR*
1940	1400	1400	331	28000+
1941	3000	4800	4000	6950
1942	4000	8600	24600+	24600
1943	6000	7000+	30000	20000
1944	9000	8000*	17000	17000

* Estimate: actual figures not available

adequate for all contingencies: when this was found not to be so, some 6-pounders were supplied, but these also proved inadequate. British tank crews were not a little irritated by being told that their guns were perfectly suitable when they knew from personal experience that the shells simply bounced off German tanks. The explanation of this unfortunate situation was available to British Technical Intelligence as early as November 1941, but was not translated into appropriate action. Most of the German tanks operating in the desert in 1942 had what was termed face-hardened frontal armour, and mounted behind it was a 75-mm gun which could penetrate 45 mm of armour at 2000 yards. In view of this, it is astonishing that the British Army managed to survive Rommel's onslaught at all, let alone defeat him, as Auchinleck did at the 1st Battle of Alamein in July 1942. When the 3rd Battle of Alamein was fought, under Montgomery's command in October 1942, the 2-pounders had been replaced by 17-pounders mounted on Sherman tanks. Of Montgomery's 1000 tanks, 300 were Shermans and their 17-pounders were as good as, if not better than, the German 88-mms.

The 88-mm gun which was mounted on Tiger tanks was, as mentioned earlier, developed by Alfred Krupp. It had originally been designed as an anti-aircraft gun, but was soon adapted to other roles as well. It proved very satisfactory as a naval gun, but its most deadly use was probably as a tank-killer. In the Western Desert the 88-mm took a terrible toll of the unwary. British and US tanks would be confronted with a few German or Italian tanks which, after a preliminary skirmish, would turn tail and run. Until they learned their lesson the hard way, British and US tanks would then scuttle after them, but the elation of victory did not last long. To their amazement – if they lived long enough to experience the emotion – the British and American tanks would find themselves confronting a well-camouflaged 88, into whose sights they had been lured. The Crusaders, the Honeys and the Grants would then 'brew-up', as tank crews describe the appalling fate of catching fire and exploding which overtook many of them. F. A. Lewis, MM, of the 10th Hussars, observed such tactics at uncomfortably close quarters at Alamein:

> It was now late afternoon, and we were surprised to see an enormous smoke-screen being put down on our right front by the artillery, and sensing that something was coming off, I took a good look round. Moving up behind us came 'C' squadron, who headed straight for the smoke cloud, moving in line ahead the better to negotiate the narrow passage open to them. The air became alive with wireless messages, one of them to watch 'C' squadron closely and put down covering fire where necessary. We had trepidation at this manoeuvre, as the position they were about to attack was very strongly held and contained not a few of the famous 88s, a gun which is a veritable nightmare to any tank personnel. What I expected happened. They were allowed through the gap without a shot being fired, then, as the last tank was clear, the Boche opened up with everything he had. With every vehicle silhouetted against the smoke, and at quite short range, the heavy anti-tank guns soon began to play havoc with the squadron as they tried vainly to deploy and engage.
>
> In as many minutes three Shermans were soon blazing furiously and two had been knocked out but were not in flames, and one could plainly see the crew 'baling out' and legging it for what cover they could find from the murderous fire coming at them from both flanks and also from their front . . .

The Sherman, which tipped the balance in the desert in the Allies' favour, was their principal tank – despite the name, it was British-designed – and America produced nearly 50,000 of them. However, as heavier tanks began to appear in German and Russian forces, the Sherman lost its pre-eminence: it was found to be underarmed and outgunned. The M26 Pershing which followed mounted a 90-mm gun and was a close approximation to the German Tiger, but M26s were not produced in large numbers.

The lesson which had been learnt in the Second World War was that the tank was not just a weapon to be used in support, but one which could be used devastatingly in its own right. The German armies of 1940 had used their armoured units to break through, to encircle, to cut the opposing defence to pieces, and to create widespread chaos. They had been ably assisted in this by the German Air Force which had intimidated by dive-bombing, and harassed ground troops by low-level machine-gunning. In the desert German tanks had used hook tactics, mentioned above, but in the Soviet Union they failed because they had under-rated the problems caused by the vast distances they had to cover, and when the Russian winter closed in the Panzers were trapped in it with all their supply lines clogged.

In the post-war period the main problem faced by army planners is to decide how well tanks can perform in the nuclear age. For a few years after 1945 there was widespread doubt over whether the tank had a battlefield future at all. It made a large target and moved slowly in relation to the weapons confronting it; these now included rockets small enough to be handled by individual soldiers, but powerful enough to blast a hole in the largest tank.

However, on the nuclear battlefield it seemed that there would be few soldiers but large numbers of armoured vehicles, if only for the reason that the latter, with their thick armour, could be impervious to both blast and radiation. Obviously they could not survive if they were too close to the centre of an explosion, but there would be plenty of places where a tank could survive when an infantryman could not, and after the strike the tank would be able to enter the nuclear zone and remain unaffected by ground contamination. Ironically, it seemed that the tank might be well equipped to defeat the very weapon which, it had been predicted, would spell its doom. That at least was the theory: the facts on a nuclear battlefield might be slightly different. And by the 1980s anti-tank weapons had developed to such an extent that if a tank survived a nuclear attack there were plenty of other ways by which its career might end suddenly and dramatically.

After the Second World War Britain, having begun the war with an inadequate tank force, was determined not to be confronted with the same situation again. As a result, British designers produced a series of models which led the world. The Centurion appeared in 1945, heavily armoured but manoeuvrable, and armed with a 76-mm gun; it weighed 50 tons. Three years later the 76-mm was replaced by an 83.4-mm. The latter gun used armour-piercing ammunition, with a tungsten carbide core which enabled it to penetrate armour twice as thick as the German 88-mm could ever manage. In 1958 the Centurion was once more up-gunned, this time with a 105-mm. The Centurion was bought by many other countries, including neutral ones such as Sweden and Switzerland, and was used in battle by the Israeli Army.

While the Centurion was still in service, its successor, the Conqueror, was introduced. The Conqueror was 15 tons heavier than the Centurion and, although it carried an even larger gun (120-mm), this did not compensate for its extra weight; the Conqueror was taken out of service in 1958, when the Centurion was up-gunned. The Chieftain, the next arrival in this impressive parade of main battle tanks (MBTs), as they were called, had some teething troubles with engine and transmission and did not go into service with the British Army until 1967. A 54-tonner, it is armed with a 120-mm gun, two 7.62-machine guns, a laser rangefinder and six smoke dischargers. It has a speed of 30 mph and a range of 280 miles, and can cross a 10-foot-wide trench. It has a crew of four; the driver is virtually horizontal in his position at the front, and this feature enables the height to be kept low. Chieftains have been sold to several Middle Eastern countries where they have acquired other names, such as Khalid and Shir. The later models have the redoubtable Chobham armour (so-called from being developed at Chobham, Surrey).

The arrival of a Chieftain on a battlefield is not likely to pass unnoticed: it fires a variety of ammunition ranging from 7.62 to HESH (high-explosive squash head) and APDS (armour-piercing discarding sabot). HESH is a particularly unpleasant, though effective, weapon which compresses itself on to the target and creates a shock wave which causes pieces of metal to break off inside an opposing tank's armour and fly around in the interior. The APDS is an ingenious device consisting of a sleeve which splits off the moment the high-velocity round has left the barrel of the gun. The Chieftain has full night vision including an infra-red searchlight. It is air-conditioned to enable it to travel through contaminated zones: the air it takes in is purified in a special pack.

The Challenger is heavier, at 60 tons, has the new high-pressure rifled 120-mm guns and can reach a speed of 35 mph. It has Chobham armour and laser sights, and shows greater versatility over uneven ground. It carries a crew of four, has been very impressive on trials and will probably remain in service until at least the 1990s.

The West German contribution to tank development is the Leopard I, which was first produced in 1965 and has been updated regularly since. The Leopard I A4, which was supplied to the German Army, was the last of a series of prototypes. In the latest models its equipment includes a stabilization system for the main armament, improved tracks and a thermal sleeve for the gun barrel.

The Leopard, which carries a crew of four, has one 105-mm gun, two 7.62 machine guns and eight smoke dischargers; it weighs 42,400 kg. The engine and transmission are at the rear of the hull, which makes it possible for the engine to be taken out in less than thirty minutes. The Leopard has a range of 373 miles and a road speed of 40 mph; its speed over rough country can, however, produce 'seasickness' problems for its crew. Among its other accomplishments are the ability to cross a 3-metre-wide trench, ascend a 60 per cent gradient, and ford rivers up to a depth of 2.25 metres without a Schnorkel, and 4 metres with one.

The Leopard 2 travels even faster, at 45 mph; its other characteristics are much like its younger brother's except that it carries a 120-mm gun and is also considerably heavier at 55,000 kg. It has Chobham armour and its engine can be removed in even less time than that of the Leopard I.

The American M1 Abrams first went into service in 1982, and has proved very popular and successful. By 1984 some 2000 had been manufactured, and this figure is expected to increase to 7000 before the 1990s. The tank is an interesting example of NATO co-operation: it has British Chobham armour and a 120-mm West German XM 256 gun; it carries twelve British-designed smoke dischargers and a Belgian 7.62 machine gun. The Abrams is slower than the Leopard at 30 mph, but has the advantage of rapid acceleration (6–20 mph in seven seconds). The gas turbine engines can run on a variety of fuels, for example petrol, diesel and jet, and like its contemporaries can be removed in less than thirty minutes. The Abrams is capable of being operated at night and under the exacting conditions of nuclear, bacteriological or chemical warfare.

A tank with a variety of uses is the American M48 A5, developed from a range of M tanks which went into service with the US Army in 1953. It is armed with a 105-mm gun, and a coaxial .3 machine gun. Many improvements have taken place since the first M48 was developed. The M48 A5 has a road speed of 30 mph and a range of nearly 300 miles. Its other performance figures are marginally below those of its contemporaries, but it is easily adapted for flame-throwing and bridge-laying purposes.

The successor to the M48 range, which initially supplemented rather than replaced the earlier models, was the M60, which first went into production in 1960. It was a bigger and heavier tank but had the same 105-mm gun, a coaxial 7.62 machine gun and a .5 anti-aircraft gun. A later version included many improvements, such as laser rangefinders, better tracks and infra-red driving lights. The M60 range has now been sold to many other armies. In the first half of the present century the export of armaments was considered by the general public to be a thoroughly evil and unpatriotic practice: today everyone is conditioned to seeing arms sales as valuable contributions to their country's export trade. Attitudes have indeed changed, and now it is not unknown for spares to be manufactured for the weapons of potentially hostile countries. The spirit of Alfred Krupp lives on.

As the Israelis have been engaged in several wars during the last few decades, it is not surprising that they have developed an excellent MBT of their own in addition to buying tanks from other countries, such as Centurions from Britain and M48s and M60s from America. Their Merkava (Mk 2) is a large tank at 60,000 kg, has a slightly slower-than-average road speed at 28 mph and a lesser range at 250 miles, but is adaptable to conditions which range from desert sand to mountain passes. It has an American engine, a British 105-mm gun, and a 7.62 machine gun as well as a 7.62 anti-aircraft gun. The Merkava is constantly updated and now has all the sophistication of contemporary MBTs.

Italy, with the OF 40, and Sweden, with the Stridsvagn 103, have useful tanks. Italian factories also manufacture Leopards and M60s. The Swedish tank has an unusually low profile and a slightly faster speed than its contemporaries. It also has a flotation system and can average 4 mph on water. It is, of course, a tank specially adapted to the Swedish terrain, which is somewhat different from that of central Europe.

The Soviet T 64 has a crew of three, and is smaller at 42,000 kg but faster at 43 mph (in the later models). Its range is 280 miles. For armament it has the 125-mm smooth-bore gun, one coaxial 7.62 machine gun, and one 14.5 anti-aircraft gun. It is protected by armour which bears a close resemblance to British Chobham design. The T 72 has a 125-mm gun and an integrated fire control system (IFCS). This, and an automatic loader, enable the crew to be one fewer than in NATO MBTs. It fires the standard range of shells: APDS, HE and HEAT (high energy anti-tank). The T 80 is said to have better armour than the T 72 and to fire a more powerful shell. Perhaps the most notable facts about Soviet tanks are the enormous numbers that the Soviet Army possesses and the speed at which Soviet industry can produce them. However, impressive though Soviet tanks are, they are not immune to the faults which all complicated pieces of machinery can so easily develop, and in the T 72 the transmission is very troublesome. It is also worth bearing in mind that massive tank armies are liable to precisely the same traffic congestion problems which confound less lethal vehicles. They also need large quantities of fuel, ammunition, spares and highly trained operators.

Other countries which possess their 'own' tanks often depend heavily on designs and materials from more advanced industrial countries. In Argentina TAMs (tanque Argentino mediario) and VCIs (vehiculo combate infanteria) depend heavily on West German designs for the Marder tank and on components imported from that country. West Germany, it will be recalled, has a record of co-operation with Argentina which has also extended into the realms of the development of nuclear power.

Similarly China, with initial Soviet assistance, developed a useful MBT in the 69, but in recent years has been relying less on Soviet help than on growing Chinese industrial strength and expertise. The Chinese are prompt to take advantage of contemporary developments in Western countries, but incorporate them so efficiently that they have been able to build up a useful export trade. Among their customers are Albania, North Korea, Vietnam and the Palestine Liberation Army.

Although the Japanese Army is strictly for home defence purposes, it possesses a very useful MBT in the Type 74. The designation is the year in which it was first supplied to the Japanese Army, and the tank has been regularly updated since. It is a product of Mitsubishi, but uses a British 105-mm gun built under licence in Japan. It also carries a 7.62 coaxial machine gun, a 12.7 anti-aircraft machine gun and six smoke dischargers. It has an excellent fire control system, which includes a laser rangefinder and a ballistic computer. At 38,000 kg it is lighter than most MBTs, but this gives it a slight speed advantage, at 33 mph. It carries a crew of four and has a range of 186 miles: it can ford a depth of up to 1 metre without preparation and 2 metres with a Schnorkel. However, the 74 is about to make way for the 88 STC, which will have a 120-mm smooth-bore gun and Chobham-type armour.

The basic problem of tank design has already been mentioned. Is it better to sacrifice speed to increased firepower and better armour (Chobham armour is already considered obsolescent– its replacement will probably be 'reactive' armour, which repels rather than absorbs), or have we reached the point at which firepower is more than adequate? Is there the room, or even the necessity, for development of other functions, such as even more sophisticated fire control? Undoubtedly the fully automatic tank will soon be seen, but this may not prove to be as great a stride forward in battlefield terms as it might seem to the theorist. Inevitably the question arises as to whether most of the functions of the tank could not be performed equally well by the SP (self-propelled) gun.

In fact the modern SP gun bears a close resemblance to the tank. It does not, of course, need the tank's manoeuvrability, and its purpose is to carry a heavy gun, probably a howitzer, which can fire a wide variety of ammunition. Thus the American M 109 A2/A3, which is the latest in the M 109 range, carries a 155-mm howitzer and a 12.7-mm Browning anti-aircraft gun. The 'how' can fire everything from chemical to high explosive, including tactical nuclear rounds. The rate of fire averages one round a minute, which provides problems for the logistical support, but the gun's all-round efficiency has made it widely popular. It has a road speed of 35 mph and a range of 217 miles, it can climb a 60 per cent gradient and cross a 6-foot trench. It can negotiate rivers up to 6 feet deep, and has infra-red driving lights. Because it is larger than many tanks it can carry a crew of six (commander, driver, gunner and ammunition handlers). Unlike the tank, an SP gun does not expect to find itself engaged in close-quarter combat, and is less likely to have problems with anti-tank guns than with enemy aircraft. Closer concerns are the possibility of fumes becoming more than enough for the extractor to manage, or a fault developing in the recoil system.

Now coming into service is the even more formidable US M110 A2, which has a better all-round performance – a range of 325 miles, and the ability to cross wider or steeper obstacles. The 203-mm howitzer can fire a wide variety of ammunition, some of which is rocket-assisted and some of which has nuclear warheads. A useful addition to the all-round firepower is the stock of some two hundred M 42 grenades which can be fired a distance of over 25 miles. The crew of this all-purpose weapon is large – thirteen – but eight of them are in the tracked weapon carrier which follows it.

The Soviet army's range of SP howitzers extends from the 122-mm to the 203-mm. The 152-mm has a crew of six, a 7.66 anti-aircraft machine gun and a road speed of 35 mph. The general performance of these tanks in climbing, obstacle-crossing and so on is the same as for NATO tanks. The rate of fire is comparatively slow, averaging one round a minute. Little is yet known about the successes and failures of the 203-mm.

The Soviets also have access to the Czech 152-mm Dana, which has a Soviet gun that is better protected than the majority of SP guns. It is light, at 23,000 kg, and has a road speed of 50 mph and a range of 600 miles. The greater speed and range are provided by dispensing with tracks and using wheels instead, but overall the Dana looks somewhat vulnerable. It has a crew of six.

Meanwhile Italy has developed a useful weapon in the Palmaria. Armed with a 155-mm howitzer and a 7.62 anti-aircraft machine gun, and weighing 46,000 kg, the Palmaria can reach a speed of 37 mph over a 250-mile range. It has a crew of five. Some two hundred have been supplied to Libya.

The French AMX-GCT is a very useful all-round weapon which carries a 155-mm howitzer, a 7.62-mm anti-aircraft machine gun, four smoke dischargers and a crew of four. With a weight of 42,000 kg it can range over 280 miles at speeds of up to 37 mph. It is fully provided with NBC (nuclear bacteriological chemical protection) and night-vision equipment and can go through water up to 7 feet deep without additional protection. The GCT (grande cadence de tir) is considered by some to be rather expensive and over-elaborate for an SP gun, but has the advantage, through skilful design, of being able to fire eight rounds very rapidly. This factor, in these days of rapidly acting enemy-locating radar, is an advantage not to be underestimated.

The latest development in SP guns is the – as yet unproven – 155-mm SP-70. A product of NATO co-operation, it incorporates elements from Britain, West Germany and Italy. At this stage there is reluctance to disclose details of its performance, although its 155-mm howitzer is expected to have a much higher rate of fire than present-day production models. Trials have shown the SP-70's all-round performance to be impressive (but it has now been abandoned).

SP tank destroyers have slightly different characteristics, though belonging to the same family. The Soviet ASU-85 carries an 85-mm gun, a 7.62 machine gun coaxially mounted, and a 12.7-mm anti-aircraft machine gun. It is light enough, at 15,500 kg, for airborne use (which is its main employment), has a road speed of 28 mph and a range of 160 miles. It is capable of climbing 70 per cent gradients and crossing 1-metre obstacles. It holds a variety of ammunition, and is equipped with an NBC system.

Germany, Austria and Sweden have also produced formidable tank destroyers. The Swedish IV-91 has a 90-mm gun, and 7.62 coaxial machine gun, a 7.62 anti-aircraft machine gun and twelve smoke dischargers. It has a road speed of 40 mph and a range of 350 miles, and can manage 4 mph on water. All-round protection is good, and there is provision for a more powerful gun in the latest designs.

The German Jp2 (in full Jagdpanzer Kanone, meaning tank-hunting gun) and its companion the Rakete both have the speed (44 mph) and range (250 miles) to carry their 90-mm gun and supporting machine guns to the point where they will be most effective. They are highly mobile, have a low silhouette, with 55 mm of armour at the front, and carry a number of anti-tank missiles of the type to be discussed later.

Accompanying the SP guns on the battlefield is a considerable variety of normally towed artillery. The US Army has the M 198, which is a towed field 155-mm howitzer with a range of close on 30 miles (with a rocket-assisted round). It can be towed by several different types of vehicle or carried to the appropriate location by helicopter. The rate of fire is four rounds a minute for three minutes, which then stabilizes to two per minute. This gun is used by many other countries.

France has a gun of similar calibre, though with a smaller crew (eight instead of ten). The TR is slightly heavier than its American contemporary, but achieves a marginally longer range.

Although these seem to be adequate for NATO needs, there is also a combined product from West Germany, Italy and Britain which is known as the 155-mm howitzer FH-70. It has a crew of eight, a light combat weight at 9300 kg, and considerable versatility; its performance is equivalent to those described above. The principal virtue of the FH-70 is its reliability.

This can also be said of the British 105-mm gun, which has a crew of four, a weight of under 2000 kg and a range of some 1.72 km. It is the replacement weapon for the 105-mm Pack Howitzer. A valuable feature of the 105-mm is its two barrels, one of which takes British Abbot ammunition and the other the American M1. Removing or replacing the barrel takes less than thirty minutes.

The Soviet Army has a useful weapon in the 122-mm D-30 towed field howitzer. This has a crew of seven, a combat weight of 3000 kg and a range of 1.5 km. It fires a variety of ammunition – the conventional and chemical range and also a HEAT shell. Like certain other Soviet weapons its general reliability has made it a popular weapon in the Warsaw Pact armies.

Apart from the problems caused to tanks by the presence of tank destroyers on the battlefield, a range of infantry anti-tank weapons also poses a formidable threat. One of the most successful is the Milan – the 'bunker buster', as it has been nicknamed. The Milan (missile d'infanterie léger anti-char) was a joint development by West Germany and France and has been produced in large numbers for NATO and other armies; the British Army uses Milan produced by British Aerospace under licence. The Milan is a wire-guided missile propelled by a two-stage rocket motor, and works on the SACLOS (semi-automatic command line of sight) guidance system. It can be fired at ranges from 27 to 2000 metres; at the latter it is travelling at 200 metres a second. Two of its 6.65-kg missiles can be carried by an infantryman. His companion will be carrying a firing post, tripod mount, optical sight and infra-red tracker. The rocket motors increase the speed of the projectile until it reaches maximum speed at 2000 metres. When it reaches its objective it drives into armour to a depth of 850 mm and it is 98 per cent accurate. Its reliability, lightness and accuracy have made it essential equipment in the armies of many countries; it was a decisive weapon in the Falklands campaign.

Less powerful, but by no means overshadowed, is the LAW 80, a British anti-tank weapon which is much smaller (it has a single-stage rocket) and lighter, but capable of knocking out a modern tank at a distance of 500 metres. (In the early days of the Second World War the then current anti-tank weapon virtually had to be pushed into the observation slits of a tank in order to do any damage.) The LAW 80 fires a HEAT missile, and accuracy of aim is ensured by accompanying it with tracer ammunition.

France has a similar weapon in the Strim LRAC 89 (lance-roquette anti-char de 89 mm). This has a 600-metre range and fires two main types of missile: one is the hollow charge for armour piercing, the other is a particularly unpleasant anti-personnel device which sprays 1600 small steel balls around the point of impact. The American equivalents to these weapons are the TOW 2 (tube-launched, optically tracked, wire-guided), which has twice the range of the Milan but is slightly heavier all round, and the Dragon M 47, which is slightly heavier than the British LAW 80 but has a longer range. The Russian equivalents are the AT-4 Spigot and the AT-5 Spandrel (these names originate from NATO, not from the Soviet Army). The Spigot is a wire-guided SACLOS weapon with a 2000-metre range, which is similar to, though not as powerful as, the Milan. The AT-5 Spandrel is also a wire-guided SACLOS missile. It has two-stage rocket propulsion and is usually vehicle-mounted, in groups of five. There is also the well-known RPG range, beginning with the RPG-7V, which fires HEAT and HE. An all-round improvement on the 7 is the RPG-16. Surprisingly, the RPG-18 is smaller and lighter: whether it has the power to knock out a modern MBT remains to be seen.

All these anti-tank weapons are rocket-assisted and are a great advance on the anti-tank weapons of the Second World War, which varied considerably in type and effect. At the top of the scale were guns like 25-pounders firing over open sights, and German 88s. The latter had originally been designed for anti-aircraft use. The British 3.7-inch was, in fact, a superior weapon to the 88, but for reasons of military conservatism was never employed in the anti-tank role. The Germans had other useful guns in the Pak range, of which the 3.7-mm was probably the best known. As Allied tanks became heavier and better armoured, the German guns were given a heavier calibre, eventually reaching the 88-mm Pak 43 L/71. It was very heavy by modern standards, weighing 5000 kg, but it was greatly valued because it could penetrate 226 mm of armour at 500 yards' distance.

Lower down the scale were the infantry weapons, such as the

German Panzerfaust, British PIAT, Soviet Simonov and American Bazooka. The Panzerfaust, which first appeared in 1942, was very successful and popular, with a rocket motor which directed a 3½-lb bomb at targets up to 300 metres. It increased in power as the war progressed, and its hollow charge could eventually penetrate 200 mm of armour without difficulty. The Panzerfaust was, however, difficult to aim accurately, and occasionally dangerous to the firer if he did not take care to avoid the fiery tail behind the rocket motor. The PIAT (projector infantry anti-tank) was probably the best of these weapons. It launched a 3-lb grenade to a distance of some 750 yards and was more than a match for most tanks. The Russian Degtyarov was a more conventional weapon, which fired tungsten or steel-cored bullets at the rate of ten a minute from its semi-automatic action. The Simonov was an improved version with a slightly higher performance. The Bazooka, or to give it its official name, the 2.36 rocket launcher M1, consisted of a simple tube 54 ins long. It launched a 3½-lb rocket, of which the last 8 oz was the warhead. It could fire HEAT, smoke and incendiary. Its range was 700 yards and it could penetrate 4.7 ins of armour.

There were, of course, other methods of stopping tanks in the Second World War. The Japanese, who had enjoyed a complete monopoly of tanks in the Malayan campaign, found conditions in Burma in 1943 less advantageous. Now confronted with Allied tanks, they resorted to '93' mines (which were remarkably similar to the German Tellermine) and Lunge mines, which were hollow charge heads on the end of long bamboo poles. The Lunge was pushed against the side of the enemy tank, causing the striker to set off the detonator. If the user succeeded in reaching a tank without being intercepted, he was unlikely to survive when the charge went off. However, during the Second World War Japanese infantrymen were not concerned about their own survival. Theirs was, of course, a similar motivation to that of the kamikaze pilots who were prepared to crash their aircraft on to enemy ships, but the infantryman did not even have the exhilaration of flying. His suicide mission consisted of filling a shoulder pack with some 20 lb of explosive and then diving between the tracks of an oncoming tank. Before doing so he would have set off the mechanism, which would explode the charge almost immediately. Whether he reached his planned destination or not, he would still inflict some damage on the tank. When the earlier tactics of diving under the tank became less successful, Japanese soldiers would hide in concealed foxholes which had been dug in roads for this particular purpose. If a tank went overhead, they would detonate whatever explosive they were carrying, even if it was only a hand grenade. In the event, few of them met their deaths in this way, for most were killed earlier by infantry accompanying the tanks.

Tanks were also made ineffective by extremely simple weapons, one of which was the basic Molotov cocktail. It was invented in the Spanish Civil War when Republican troops had no other means of stopping the German and Italian tanks which Hitler and Mussolini were supplying to Franco's troops with a view to testing their characteristics.

A Molotov (this was the assumed name of the Soviet Foreign Secretary at the time) cocktail was simply a glass bottle filled with petrol and tar. Occasionally diesel oil was substituted for the tar. The 'fuse' was a paraffin-soaked rag which was tied around the neck of the bottle. When the 'cocktail' was thrown against a hard object, it broke and sprayed burning petrol over the surrounding area. Often this petrol would penetrate the hatch, observation slits or even engine vent of a tank, to devastating effect. It could also damage the tracks. This easily manufactured weapon became immensely popular with partisans – and, of course, is still used by terrorists today; the British Home Guard made Molotov cocktails by the thousand in 1940 when there was a real expectation of a German invasion. The Molotov cocktail is most effective when a tank is moving through a built-up area, for it can be dropped on to it from close quarters: in the open it is less of a problem to a tank crew, for its thrower can seldom get close enough to the objective to use it without drawing attention.

Tank warfare is a phenomenon of the twentieth century, and only those who have taken part are fully aware of the strange mixture of fear and elation, discomfort and satisfaction which accompany it. The first occasion on which British tanks decisively defeated the Germans was at Alamein on 23 October 1942 and during the following days. Some forty years later T. I. L. Powell recalled:

> I was merely a trooper, a member of a tank crew in the Alamein campaign. When the time for action came it was with dramatic suddenness. I remember sitting outside my tank when the call

came: 'Tank commanders to Command Tank.' So for the first time in my life I went into battle. The tank commander, a regular army sergeant, a good-natured fellow and an inveterate borrower, who in the past had always seemed hard up, and had taken small coins from all his crew, emptied his pockets and repaid every penny he'd borrowed. I remember him saying, 'Better pay my debts, you never know.' In less than half an hour he lay body and shoulders on the tank floor, his head and brains spattered on the surrounding walls. . .

Eventually the tank I was in was knocked out. I clearly remember the driver, a phlegmatic little Yorkshireman, calling to us in the turret, 'My visor's gone. Next one we've had it.' His tone was conversational, almost casual. In a matter of minutes the tank was engulfed in flames. What surprised me was that I was not conscious of any explosion or sound, just realising I was in the middle of a fire and myself on fire. The gunner was killed, the rest of us escaped. I do remember ripping the leads of the commander's earphones from the radio set so that he could fling himself out, and hurling myself to the ground the moment his feet vanished through the hole above my head.

W. E. Bowles, MC, who was in the 10th Armoured Division at Alamein, remembers the confusion of the battlefield, the renowned 'fog of war', which could be both physical and mental:

On going forward, conforming to Monty's wishes, we found ourselves two-thirds of the way down the southern mine-cleared corridor in an unholy jam of tanks, supply trucks (wanting to get unloaded and back before daylight), Bren carriers, guns etc, in an atmosphere reminiscent of the old-fashioned London pea-souper – in those corridors the heavy traffic churned up the sand as fine dust, which filled the atmosphere and brought visibility down to three or four yards. I think this made my most nightmarish memory: the way ahead was choked with vehicles, the way back was also choked with follow-up traffic; anyone foolish enought to try a detour round a broken-down vehicle was likely to hit a mine. There was the heavy roar of a score of tank engines, frantic shouts and messages, urgent calls on the radio, telling me to 'do this' and 'go there', and into this blind maelstrom was falling a generous dose of enemy fire.

J. E. Delhanty was in a Sherman tank at Alamein. The first Shermans off the production line were to be used to equip the American Army but Roosevelt, knowing how desperate was the British need for tanks to counter the German Mark IVs, arranged for three hundred to be sent to the British Eighth Army:

> As we rolled down into the wadi [defile] we spotted eight Mark IVs, and promptly engaged them: we were in hell's delight at a battle at close quarters as we were eighty yards apart. We received one on the turret; the one that sent it received a couple from me straight in the guts: we got all the crew with our Browning machine-gun as they baled out. I polished off another but then we were hit with a shell from a squeeze gun. The barrel only fires forty rounds, then it is kaput. The diameter of the shell is two inches and leaves the barrel measuring only one and a half inches. It came straight inside the tank just behind my back and set the anti-personnel ammunition on fire. It was like breathing hell-fire, so we decided to bale out as we had ten hundredweight of high explosive in stock. As I rolled off the deck I went up in the air as Jerry dropped a few high explosives around us. I felt a knock in my back. I rolled into a small depression and I was sure my legs had gone, then I had terrific pins and needles and found they were in working order. I heard Blackie, our radio op [whose feet had been blown off] screaming from the side of the tank, so I crawled back to him and pulled him away from the tank as I knew she could blow up at any second. We were only fifteen feet away when the turret flew straight up into the air. It appeared to turn over slowly and then dropped back on the tank.

During the Second World War tanks were often required to perform under conditions which earlier would have seemed impossible. On D-Day and later in the North-west Europe campaign one division, the 79th, included what the rest of the Army likes to describe as 'the funnies'. It included DD tanks (duplex drive) which were amphibious, 'crocodile' tanks which carried flame throwers, 'crab' tanks which were armed with flails to detonate the mines in front of them, and AVREs (assault vehicles Royal Engineers) which could either lay down bridges or carry huge 'petards' for demolishing obstacles. They were invaluable in clearing the beaches of obstacles.

When R. Cadogan landed on D-Day he was nineteen; later he wrote down his impressions of his experience of serving in a DD tank:

We made good progress, the tank behaving magnificently in a sea which would have daunted many a larger craft, leave alone a tank. Never in our wildest hopes had we thought they would stand up in such a sea.

By now the shore was about two hundred yards away and some of the tanks had already touched down, deflating their canvases as they beached. We were close behind a tank which had the misfortune to run onto a mine. The canvas collapsed and water poured through a jagged hole in it. The crew, as calmly as though they were on an exercise, climbed out and on to the deck of another tank which had pulled alongside.

We were not so lucky. We beached, but by a stroke of ill-luck hit an anti-tank obstruction. The driver slammed the tank into reverse, and we backed a couple of yards, but too far. With our canvas already deflated, the sea just poured in, naturally stopping the engine. So there we were – finished before even starting. Still, as our orders were to destroy any houses left standing in the village we just stayed put and kept the gun firing.

Major Peter Selerie went in with the Sherwood Rangers Yeomanry:

On beaching we solved the mystery of the black patches in the aerial photographs supplied by the RAF. They were boggy patches, and already one or two tanks had fallen foul of them. I ordered the drivers to engage emergency low gear and we managed to negotiate this hazard.

There was quite a shambles on the beach. The flail tanks which were supposed to clear the main exit lane through the minefield had been knocked out. Fortunately, an extremely brave Sapper officer offered to clear a secondary lane manually. We gave him and his small party covering fire. Out of the squadron's nineteen original tanks, only five were still mobile and I ordered the first two, closely followed by the remaining three, down the lane. We were engaged by an anti-tank gun firing from a concrete casemate. He scored a hit on the leading tank, destroying the gun mantlet on the front of the turret. With all five tank guns however we managed to silence him. We then turned on to the road leading to Le Hamel. Pausing in the outskirts to check enemy defence, we were overtaken by one of the AVRE Churchill tanks armed with a petard that looked like a short and very wicked piece of drainpipe sticking out of the turret. . . .

When we entered Le Hamel it appeared that the main enemy fire was coming from a tall, many-storeyed house. I ordered the Churchill tank forward to demolish the house with the petard, which had a very short range. Maximum covering fire was given by the Sherman tanks' 77 mms and machine-guns. The petard fired and something like a small flying dustbin hit the house just above the front door. It collapsed like a pack of cards, spilling the defenders with their machine-guns, anti-tank weapons, and an avalanche of bricks into the courtyard.

On 9th June (D + 3) there was heavy fighting and the enemy seemed to have succeeded in surrounding our position. We sustained a number of casualties, including the poet, Keith Douglas, who was killed. It became increasingly obvious that our 75-mm gun would not penetrate the frontal armour of the German Mark VI (Tiger) or the Mark V (Panther) tanks. It was exceedingly difficult to get on their flank and fire at the side armour. We heard that some Shermans armed with the British 17-pounder had arrived in Normandy. This weapon possessed far superior powers of penetration and we sent urgent messages back requesting the new tanks. To help us in our offensive it was therefore necessary to have maximum artillery so that a tremendous 'stonk' could be put down on the enemy armour. It is interesting to note that the British handling of artillery had evolved in a superior way to that of the Germans, who were great lovers of little pockets of various arms called Battle Groups, which led to fragmentation. The British concept was that the whole of the fire power, for example a corps' artillery plus that of formations on the flank, could be brought down on a small sector of the front.

Lieutenant-Colonel Peter Sutton, of the Royal Tank Regiment, had the onerous responsibility of commanding a squadron of 'funnies' at that time:

> I was in command of a breaching squadron of eighteen assorted specialised armour, flail tanks for mine-destroying, assault vehicles Royal Engineers for destroying concrete pill-boxes, laying mats across mud, and even one tank carrying a 30-foot bridge. We were travelling in three tank landing craft, and a similar squadron, under the command of Major Tim Thompstone, Royal Engineers, was in another three.

Fortunately for this squadron of 'funnies', their landing was comparatively uneventful. Their turn for stern resistance was to come later. However Major Tim Wheway, MC, ran into trouble immediately:

> At 0720 the Landing Craft Tanks go full speed ahead and it is a race for the shore. We land at 0725 hours and the impact nearly shoots the tanks through the doors. The flails stream out in three feet of water, followed by the AVREs. We are met by terrific shell, mortar, 88 and 75 Anti-Personnel, and small arms fire, at three hundred yards range. The LCT with our commanding officer, Lt-Col Cox, 5th A.R.E. is hit. Leading flail on the ship manages to get off but Corporal Brotherton is killed and the crew of the second is wounded. Colonel Cox is killed. Several tanks are hit as the landing craft doors go down. Mines are sighted on top of the wooden beach obstacles. We go down as far as possible in the water to be able to use our guns effectively and then open fire on concrete gun emplacements, houses, and dug-in infantry. Tanks are brewing-up right and left. We then proceed flailing out gaps, but no mines are encountered so speed up and get within fifty yards of our gapping [clearing lanes through the minefields] places and fire right into the slots of gun emplacements. One flail tank strikes a sunken obstacle with a mine on it and the bottom is blown in. Lieutenant Robertson's has a direct hit on the flail's arms, knocking them off. Lieutenant Allen's has three 88 mm AP straight through into the turret and all but one of the crew are killed. Wounded and burnt personnel succeed in getting into the sea and are picked up by a Landing Craft Ambulance. Corporal Agnew's tank has three 88 mm AP through the engine, and brews up. Trooper Jennings is wounded in getting the driver's hatch undone. An 88 AP goes straight into the front of Sergeant

> Cochran's tank, killing the operator, Trooper Kemp, and wounding Sergeant Cochran and Trooper McKinnon. Some flails now start gapping and the East Yorkshires and South Lancashires are now streaming up the beach covered by fire from the beach-clearing flail-tanks. The AVREs follow the flails and the bridging AVREs drop their bridges, but the crews, who jump out to make them fast, are killed or wounded in doing so, and the tanks receive direct hits and are brewed up. German soldiers rush from the houses shouting and firing as they come and soon the beach is strewn with dead and wounded of our own and enemy troops.
>
> The beach clearance flails are now waiting for the 629 REs to assist them in clearing the beach. Captain Wheway and Lieutenant Sadler get out of their tanks and attempt to contact them. Shells and mortars are falling thick and fast but no one realises the danger of them until Trooper Hogg is killed by the side of his tank. A lieutenant of 629 is eventually found and he states he is the only officer left and their casualties are so heavy they cannot assist us, so clearance flails proceed up the beach and commence gapping defences where they can to clear the congestion on the beach. Sergeant Turner and Corporal Aird returned to the beach after having successfully made their gaps and flailed their laterals, but small arms and shell fire is still intense and both Sergeant Turner and Corporal Aird are killed by snipers' bullets. After two hours fierce fighting enemy resistance is wiped out, lanes are made, the laterals are clear, and the surviving flails are back on the shore.

In those few paragraphs Tim Wheway describes, from his own experience, what happens when weapons meet weapons. It is, perhaps, easy to forget that the purpose of guns is to kill people and the function of tanks is to take them to places where this may easily be done. The cost to each side is human lives. The battlefield brings out the finest human virtues in the most appalling circumstances. Weapons which incorporate the ingenious skills of designers and manufacturers are handled by men (or women, under certain conditions) with courage and steadfastness which at times seem almost unbelievable. At the end of each phase of a battle, when the dead are being buried and the next of kin are being informed, the survivors can but think that there must be a better way to run a world than for the young people of civilized countries to blow each other to pieces with sophisticated machinery.

8

Rockets and Missiles

The rocket, which seems to us the most modern of weapons, has in fact been an important means of waging war for over a thousand years. The pioneers of this weapon – and of many others – seem to have been the Chinese. They are known to have discovered gunpowder by the ninth century, but initially used it only to make fireworks; their delight in pyrotechnics persists to this day. Eventually the Chinese came to realize that this new discovery could also be used to make an intimidating weapon of war. The first adaptation of the rocket for war was comparatively crude, for it was merely attached to an arrow or a spear; soon, however, its possibilities as an incendiary weapon (literally a form of firepower) were being exploited. The new discovery travelled across Europe with the Mongols in the mid-thirteenth century, adding to the terror already created by those apparently invincible barbarians. By the fifteenth century rockets were in use in Italy and Spain. It may not, however, be entirely fair to place the responsibility for the introduction of rocket warfare on the Chinese: as seen earlier, when nations reach a certain stage of scientific development they seem to make discoveries for good or bad concurrently. Early in their development rockets were adapted for signalling. It seems, however, that there were more theorists writing about the use of rockets than experts in their practical use. As is only too well known today from the failure of a number of space rockets, a major problem is keeping a rocket on course: unless it travels accurately to the appointed target or destination, it can easily become more destructive to the user than to his opponent.

Nevertheless, by the eighteenth century perseverance in the use of the new weapon was beginning to foreshadow modern developments. To the surprise and consternation of Britain, the next step was taken in India, which was not at that time thought to be very advanced in military science. In 1792 and 1799 Hyder Ali and his son Tippoo employed a force of five thousand rocket troops. The combustion chamber in their rockets was an integral part of the missile. When the rocket, which was stabilized by being attached to a long bamboo stick, reached the target it was likely to ricochet along the ground, causing considerable disturbance among the opposing cavalry, which was usually British.

The next stage was the Congreve rocket. Sir William Congreve was a British artillery officer who in 1805 built a missile which was 40 ins long, stabilized by a stick 16 feet long, and had a range of 2000 yards. This very practical weapon was used in various battles of the Napoleonic Wars, notably at Boulogne, Copenhagen and Danzig. In 1814 Congreve's rockets were used when the British Army attacked Fort McHenry, in Maryland, during the 1812–14 war between Britain and the United States. It was the appearance of Congreve's rockets that caused Francis Scott Key, author of 'The Star Spangled Banner', to use the words: 'The rockets' red glare, the bombs bursting in air'. By this time the rockets had become sophisticated enough to have either an explosive or an incendiary warhead, and by using various lengths of fuses it was possible to explode the warhead in the air when required. There was no doubt about the formidable nature of these weapons, which eventually acquired a range of 2 miles. Although the performance of Congreve's rockets was soon outclassed by developments in normal artillery in the early nineteenth century, his design was copied and improved upon by many European countries. A glimpse into the future was provided by a Swedish development which had no stabilizing stick but could maintain an even course by being mounted on a delta (triangular) wing. It was fired from a hand-held launcher of the type which would become all too well known over a century later.

Less renowned than Congreve but equally important was William Hale, a British engineer who in the 1840s successfully replaced the guide stick by inserting vents at appropriate angles in the combustion chamber, thus imparting a spin to the rocket in flight. However, even Hale's rockets were unable to compete with the latest developments in continental ordnance. Surprisingly they found most favour in the

Russian armies, which considered them ideal for mountain warfare. Two thousand were made for the US Army in the Mexican war of 1846–8, and carried a 16-lb warhead, but were not considered as effective as other forms of artillery. This did not prevent them being tried again in the American Civil War of 1861–5, but here again they found little favour.

However, like many other instruments which were originally invented for military use, the rocket had notable successes when applied to more humane purposes. The nineteenth century had an infamous reputation for the number of ships which were lost close to the shore, often sinking with all their crew and passengers, because it was impossible to launch lifeboats into the dangerous sea. Although Congreve's rockets were not the first to be used for the purpose, they were one of the most effective means of shooting a line from ship to shore, or vice versa. Once a line had made the essential link, it was comparatively easy to use it to convey a stronger line and ultimately the survivors of the shipwreck. The heyday of this form of rescue was in the days of sailing ships, which could all too easily be driven on to rocks: the arrival of steamships and oil-burning ships did not completely eradicate the danger of shipwreck – which indeed can still occur – but it made rocket stations much less necessary. Rocket propulsion had many other applications, too. The rocket-fired harpoon dramatically increased the toll on whales. And even before the end of the nineteenth century there was much theorizing about using rocket propulsion to power various forms of aircraft: it was even suggested that if a more appropriate fuel than gunpowder, perhaps a liquid, could be devised, rockets might be able to soar above the earth's atmosphere. These concepts were not the fantasy creations of early science fiction writers but were based on the theories of well-qualified scientists and mathematicians.

Although rockets were seldom used in warfare in the first few decades of the twentieth century, research and experiments produced impressive results. Like many other developments in military science they were neglected by senior officers and were unknown to the general public until that memorable day, 13 June 1944 – one week after the successful invasion of Normandy – when the first V1 fell on London. This 'flying bomb' came as a great shock to the citizens of London, who had already been battered remorselessly by the Luftwaffe for several years and now hoped that their ordeal was virtually over; the rocket was spoken of as a 'revolutionary new discovery', 'Hitler's secret weapon', and 'a new method of waging war'. In fact the arrival of the V1 should not have taken anyone by surprise, for all its characteristics had been predictable for over fifty years. A major contribution was made by Sweden, both in the design of the motor and in the type of fuel used. The Swedes had estimated that any form of aircraft which threatened their country would be outside the range of conventional anti-aircraft weapons, but that rockets powered by liquid fuels, which would be able to accelerate once in flight, would make short work of both distance and evasive tactics. Not surprisingly, the forward-looking Krupp bought both the patents and the prototype in 1909.

However, the next stride forward came not from Germany but from the USA, where a brilliant designer named R. H. Goddard produced a greatly improved rocket motor with a tapered nozzle. He also produced a prototype for a long-distance, high-altitude rocket which would proceed by stages, on the principle of present-day rockets. As there did not seem to be any obvious application for these large rockets during the First World War, Goddard turned his attention to rockets which could be used on the battlefields of the day. He did not care greatly for this diversion from his principal interest – which was now the possibility of space exploration – but he did succeed in producing a series of prototypes for hand-launched rockets. Though they were too late for the First World War, his experiments proved of great value when a device for stopping tanks was urgently required in the Second: it is to Goddard that the US Army owes the development of the bazooka, described in Chapter 7.

After the end of the First World War he returned to his cherished large rockets, now powered by a mixture of liquid oxygen and petrol. His first experimental rocket, which was launched in 1926, proved a disappointment, for it did not travel more than a few yards before it crashed: however, those few yards had shown that the basic principle was correct and that this was undoubtedly the weapon of the future.

Further and more successful experiments were carried out, using gyroscopic stabilizers, but progress was slow and setbacks were often encountered. Goddard concluded various experiments in New Mexico and in 1935 built a rocket which travelled faster than sound. However, his vision of the possibility of space – even interplanetary – travel attracted more amusement than admiration in the American press. It was all very well for H. G. Wells (whose books had originally stimulated Goddard's imagination) to write fiction about such impossibilities, but the idea of a practical scientist conducting research funded by respectable institutions caused him to be nicknamed 'Moony'. During the Second World War he was in his sixties, and the only employment he could find was work on the application of jet boosters for seaplanes. He died in 1945, by which time Germany might even have defeated Britain had the V2s been developed a little earlier. The V2 was powered by the petrol and oxygen fuel which Goddard had invented in 1926, and within a few years of his death all his dreams had come true. America had landed the first men on the moon and several countries were conducting experiments which Goddard, long before, had made possible. The end of the First World War had come before Goddard's prototype bazookas could be tried in battle, and the Second had finished before the US military establishment realized that Goddard's discoveries would rule the battlefields of the future.

Goddard was not alone in conducting experiments with rockets, although he seems both the most undervalued and the most successful of the pioneers. Elmer and Lawrence Sperry experimented with a small pilotless aircraft which flew successfully in the United States in 1917, but was too late to be used before the war ended. It was to be kept on course by gyroscope and was programmed to deliver an explosive warhead to a chosen target. Another inventor who worked on pilotless aircraft was the British Professor A. M. Low. Low was undoubtedly a genius but his ideas, which often featured in Sunday newspapers, were considered too far-reaching, and thus impractical, by conservative military thinkers.

Other countries were now beginning to give serious consideration to the idea of rocket propulsion. In Germany in 1929 Hermann Oberth, who was not even a military designer but merely a schoolteacher, published a paper on space travel which was subsequently considered to be some thirty years ahead of its time in its concepts. In France Robert Esnault-Pelterie, an experienced test pilot, added his contribution and gave us the word 'astronaut'. The Soviets were by no means behind but conducted their experiments with increasing secrecy. The best-funded experiments were undoubtedly those conducted by Wernher von Braun, a member of the German Rocket Society who had successfully approached the

German Army for backing. Braun's career in this field began when he was appointed leader of a small technical group; within a few years he was guiding a huge team of engineers and technicians conducting experiments on a desolate island off the Baltic coast. The island was Usedom, and the name of the site was Peenemünde.

During the Second World War Allied Intelligence learned that Peenemünde had a special importance to German military plans and began to take photographs. A study of them suggested that the attention of Allied bombers might at least help to delay production of what could turn out to be an extremely dangerous German weapon. By this time, however, German experiments in the manufacture of military rockets had spread well beyond the concept of long-distance rockets of the V1 and V2 type. The Panzerfaust, for knocking out enemy tanks, has already been described; a more formidable weapon, greatly detested by its opponents, was the Nebelwerfer, a device with up to eight barrels which launched 80-lb rockets at the rate of one a minute to a range of 6000 yards. These were, of course, overshadowed by the notorious long-range missiles. The pilotless V1 was powered by a pulse-jet engine and carried a warhead of 2200 lb. It was an awesome weapon because those in the target area would hear its engine as it approached, then realize it had stopped and that a shattering explosion was imminent. Eventually the V1, bad though it was, seemed less terrifying than its successor, the V2 rocket, the first of which landed on London on 8 September 1944.

Nevertheless the V1 did enormous damage in Britain. It is easy to be wise after the event, but the fact that British Intelligence was well aware that German scientists were working on pilotless aircraft and rockets before 1939, and that before and during the war the RAF were using the Queen Bee, a pilotless target aircraft with a range of 300 miles, suggests that British military planners should have considered the possibility of offensive flying bombs more seriously than they did. When the V1s began to arrive, the indiscriminate manner in which they found their targets added to their morale-destroying power. A mere five days after the first one had landed, the public was shocked to hear that the Guards Chapel in Wellington Barracks had been hit by a flying bomb during Sunday morning service: 119 people were killed and a further 102 seriously injured. Within a few hundred yards of the Guards Chapel lay Buckingham Palace, most of the important government offices and the Houses of Parliament: this was really striking at the heart of the capital. The number killed by V1s elsewhere during that week brought the total to over 500, and the destruction of buildings was enormous. As the weeks went by horrifying stories were heard about V1s landing on hospitals and crowded restaurants. The fact that the devastation was caused by a robot bomb made it seem worse for many people; being bombed by an aircraft with a pilot was bad enough, but one could at least feel that the pilot was putting his own life at risk while doing so; to be destroyed by a V1 seemed so coldly impersonal that, in a strange way, it was more frightening. Subsequently it was revealed that many V1s had landed far from their appointed destinations. A large number had been targeted on Southampton, to interfere with the supply lines for the Normandy invasion, but all had fallen miles away. The German Army was never able to obtain a correct account of the performance of the V1s, for German spies who had been captured, 'turned' and were now under British control, sent back totally misleading reports about their accuracy. However, in spite of successful counter-measures in the shape of fighter aircraft and anti-aircraft guns, the doodlebugs, as they were nicknamed, continued their path of destruction in many parts of the country, notably East Anglia, until March 1945.

The V2 (Vergeltungswaffen = vengeance weapon) carried an even heavier warhead, and weighed in all 13½ tons (the explosive weighed approximately 1 ton). It had a range of 200 miles. Just over one thousand were fired at London: three thousand were aimed at targets elsewhere. The V2 travelled in a vast arc reaching a height of 70 miles, taken on its lethal way by a propellant-supply system. Its arrival without audible warning was due to its moving faster than the speed of the sound that its progress created.

Although the German rockets did not come in time to be a war-winning weapon, the fact that their scientists and engineers had accumulated a vast range of information about the potential of rocketry caused these men to be eagerly sought after in the days following the German defeat. The Soviets captured Peenemünde itself, but found that the Germans had been careful to destroy everything of potential value to their conquerors; the Russians, however, managed to take their share of German rocket engineers and scientists. The USA did slightly better, for as well as obtaining Wernher von Braun it acquired the contents of a huge underground rocket factory: with these, American scientists were able to conduct many useful experiments. Even so, America placed more reliance on long-range bombers than on strategic missiles. Rather than concentrate on developing intercontinental missiles, the United States preferred to produce battlefield weapons such as the Corporal, with a 75-mile range, or the Pershing with 400; the latter was to have a long career, and is still in use to this day.

In the 1950s various developments contributed to make rockets infinitely more powerful and dangerous. The fact that, after the discovery of the hydrogen bomb, they could be given a much more powerful warhead, and that guidance systems improved dramatically, meant that intercontinental ballistic missiles could not merely travel a quarter of the way around the world but could also land with a fair degree of accuracy when they reached their allotted destinations. Atlas was first fired in 1958; Titan in 1959. Minuteman (described later in this chapter) followed in 1961, and was a weapon launched from 90 feet underground; it has been continuously updated. Polaris, made for launching from seaborne vessels, was tested above sea-level in 1958 and below it in 1960. There are obvious problems about carrying large quantities of liquid propellant under water, so Polaris has a solid propellant. It is actually launched by compressed air, then the first rocket motor takes over. A submarine can carry sixteen Polaris missiles. However, Polaris has now been rendered virtually obsolete by Poseidon, described later in this chapter; like Minuteman, Poseidon carries MIRV warheads. MIRV stands for multiple independently targeted re-entry vehicles, which in simple language means that an average of three nuclear warheads can be sent to different targets by a single rocket. When one visualizes batteries of American MIRVs contemplating even larger numbers of Soviet MIRVs, it is not surprising that there are enough nuclear weapons in existence to destroy every inhabitant of the world ten times over. The atomic bomb used at Hiroshima had an explosive charge equal to 20,000 tons of TNT: that in a modern 'device', as it is euphemistically known, is equal to 5 million tons.

As may be expected there is considerable variety in missiles and their launching systems. At the highest level is the MIRV against ABM, as outlined above; at lower levels come submarine-launched anti-submarine missiles (Subroc), and intermediate-range ballistic missiles (IRBMs) such as the American Thor and Jupiter, together with the Soviet range which began with Shyster, Sandal and Skean and more recently included Scud and FROG. (All of these are NATO nicknames.) Sandal was the type of missile which Khrushchev sent to Cuba in 1962, bringing the Third World War into the bounds of immediate possibility. That situation was, fortunately, resolved by John F. Kennedy's firmness and Khrushchev's possession of the power and authority to back down without losing position and prestige.

Scud (both A and B), FROG and Scrooge are all IRBMs which arouse deep suspicion in the West when Soviet military intentions and peaceful protestations are being examined. All can be moved easily and concealed in wooded country, where they escape observation by both aircraft and satellite. When therefore it is announced that they are being withdrawn from the western borders of Warsaw Pact countries such as East Germany, Western military pundits wonder whether this merely indicates that they are going to an even better hiding place a few miles distant. At the same time, countries to the east and south of the USSR speculate on whether weapons withdrawn from the west will immediately be redeployed and targeted on China, South Korea and Japan. Similar misgivings are felt by Australia and New Zealand when withdrawals of ICBMs are mentioned.

As may be judged from their successes in space exploration, Soviet progress in rocket development has been impressive. In the mid-1960s the West began to take note of the SS range of missiles. Scarp (the SS-9) is a liquid-fuelled rocket capable of leaping over NATO's early warning system. Over two hundred of these are deployed at present and each has a warhead with an explosive potential of 25 million tons of TNT.

The Russian SS range of rockets has developed steadily, though a higher number does not necessarily mean an increase in size. The SS-17 has the longest range, at 6800 miles, and carries four warheads. The SS-18 is a massive rocket with a warhead carrying ten MIRV warheads, each of which has the equivalent explosive power of 500,000 tons of TNT; its range is slightly shorter than that of the SS-17. Although there are only some three hundred SS-18s (known to be) in existence, it is thought that their explosive power, if properly targeted, could destroy over three-quarters of the United States' ICBMs, if they were used in a first-strike attack. The SS-19 is smaller, but said to be accurate and reliable. Each warhead contains six MIRVs which together mobilize 500,000 tons of high explosive, and its range is 6000 miles. The SS-20 is a subject of some contention in discussions about the reduction of armaments, for it carries a mere three MIRVs (about 30,000 tons; but has a range of 3000 miles. It is highly mobile and the Soviet Army probably has some five hundred of them. As 3000 miles puts a wide choice of target into range, the United States insists that, by any standards, this must be classed as an intercontinental ballistic missile. Equally stoutly the Soviets deny this and say it is only an intermediate-range ballistic missile.

However, the USSR has now developed new weapons in the SS range and these, though smaller, are thought to offer an even greater threat because they are more mobile still. They include the SS-21, SS-22 and SS-23, all of which are short-range (60–600 miles) missiles. NATO has very few weapons of this type – only Pershing and Lance – and even if it had more it would be unable to deploy them as effectively as the Warsaw Pact countries can. The Soviets could therefore destroy their 800 SS-20s but retain 700 of these later types. NATO is not happy with any reduction of forces which leaves the Warsaw Pact with an equivalent 'throw-weight' in another range which has not even come into the discussion. But the core of the arms control problem is not merely negotiating a balance between nuclear warheads; it is the achievement of balanced forces in Europe. While Soviet conventional forces, in the shape of tanks, guns and aircraft, vastly outnumber those of NATO, the aim of the latter must always be to retain sufficient numbers of nuclear weapons to make a Russian first-strike too hazardous an operation to be contemplated. NATO could agree to the elimination of all IRBMs, if its own forces possessed enough short-range missiles to make any Soviet invasion of western Europe too costly. NATO might even agree to remove all nuclear weapons from Europe, if Warsaw Pact conventional forces were reduced so drastically that they had no more than parity with NATO forces.

It might seem, therefore, that the problem of reducing the size and type of nuclear weapons is comparatively simple; but it is not so. Since 1917 the USSR has been ruled by a political system whose philosophy asserts that the expansion of Communism is inevitable, but that unless a Communist state is strongly armed it will be under constant attack by capitalist states. It is highly improbable that the Soviet government still believes that Western nations wish to invade the USSR and restore capitalism by force, for when Germany did in 1941 the other Western nations went to enormous lengths to help the Soviets expel the aggressor – but old beliefs die hard. The USSR would like to relax its rigid system in order to enliven its stagnant domestic economy, but has for so long had its state economy bound up with the manufacture of armaments of every type that to begin dismantling that military economy might cause chaos and disaffection. Approximately three-quarters of Soviet economic production is linked to the armaments industry. It is easier to produce more Kalashnikovs and T.80s and to give them away to countries with leanings towards Marxism than it would be to change to products which could compete freely and satisfactorily with other countries in world markets. Now, with the East and West both anxious to reduce the danger and crushing cost of vast nuclear arsenals, this relentless increase in firepower on all sides may at last be halted and reversed. Whatever progress is made is unlikely to be speedy. The Soviets cannot afford to do anything which might threaten the stability of the regime, and are well aware that, as the USSR itself is made up of a vast collection of disparate ethnic groups, political disintegration is not far below the surface.

Nevertheless, while NATO is not unmindful of the Soviet Union's problems, it is even more conscious of the fact that SS-22 missiles launched from the middle of East Germany could reach the whole of West Germany, Britain and France in their 'intermediate' flight. Similarly, SS-23s could pound to rubble all the NATO installations from Belgium to Turkey. However, after the experience of Chernobyl it seems unlikely that Soviet strategists would wish to launch an invasion of Europe using nuclear short-range weapons, as the effects would make the over-run territory uninhabitable and the wider implications would be incalculable.

The technical details of some other destructive agents remain to be discussed. The 122-mm M 1972 MRS is a multiple rocket system of Czech origin mounted on a vehicle, which makes it extremely mobile. When required, it can fire a rocket every ten minutes. FROG (free rocket over ground) is a similar weapon; though overtaken by the SS-21 it is still in use in certain units on account of its durability and reliability. Lance, MLRS and Pershing are all American missiles. Lance can deliver a variety of warheads to a point up to 75 miles away at a speed of 2000 mph. MLRS (multiple launch rocket system) has a short range of just under 20 miles but can fire, according to the type of warhead used, from two to twelve rounds a minute. This rate of fire creates considerable supply problems, but not nearly as many as are likely to be experienced by nations planning to advance over territory on to which the blanket cover of missiles is directed. Pershings are equally formidable, though of a different character. Their range is some 500 miles, over which they can deliver a nuclear warhead. Pershings are extremely accurate and have the added attraction to the user that they can penetrate deeply before exploding: the occupant of an underground bunker is liable to feel less than confident if he knows that Pershings are targeted on it.

Cruise missiles come into a slightly different category of weapon which can be launched from air, sea or land. They travel comparatively slowly, following the curve of the earth's surface, and therefore can come in under a country's radar screen. Cruise is more of a pilotless aircraft than a missile, but the fact that it can carry a nuclear warhead, which it has an excellent chance of delivering, makes it a formidable weapon in the inventory of death. Nobody is quite sure whether such weapons are tactical (for use on the local battlefield) or strategic (having a wider application). The French cruise missile is ASMP (air-sol moyenne portée), while the USSR has the SS-NX-21 and America the Tomahawk in both land and sea versions.

The publicity recently given to all forms of cruise missiles, particularly the land-based variety, is due to the fact that they represent a threat from an unknown base. In order to practise their mobility under emergency conditions cruise missiles are periodically moved into the countryside away from their normal base. Not surprisingly, the general public does not care for the thought of powerful liquid propellants moving around the countryside, whether accompanied by a nuclear warhead or not. Everyone with a fraction of common sense believes in nuclear disarmament, but an interesting feature of the present age is that protestors, being unable to interfere with the nuclear weapons of potential enemy countries, rely on trying to interfere with the readiness of the weapons deployed by their own countries.

Every country has an array of air defence weapons, which are nowadays usually rocket-propelled. Britain has Blowpipe, a man-portable SAM (surface-to-air missile). The launcher, Javelin, holds three missiles in launch tubes and incorporates an aiming unit. The range is 3 miles, and the warhead is high explosive with a proximity fuse. Blowpipe proved very successful in the Falklands War, and is now being made more powerful and generally improved. France has the Crotale, West Germany the Gepard (a self-propelled air defence gun equipped with two Oerlikon 35-mm cannon) and the United States the Patriot (MIM-104), which is altogether more powerful: its range is 37 miles, it closes on the target at some 2000 mph, and it can carry either a conventional HE or nuclear warhead. Britain also has the Rapier, which is used by many other countries too. Eight Rapier missiles are mounted beside an armoured cab, which holds a crew of three, and are ready for instant action. The latest versions have automatic laser trackers.

France and West Germany produced the Roland jointly and it is now in service with both countries as well as many others. The Argentines used a Roland in the Falklands War, but with limited success. The USA took a small number when it was first produced, but subsequently preferred the Sergeant-York (M 998 DIVAD) and the Stinger. The former is a self-propelled gun with a crew of three. The gun, or rather guns, are two 40-mm L/70 Bofors which give a combined rate of fire of 620 rounds a minute, automatically targeted. Veterans of the Second World War will always remember the steady, rhythmic thumping of the Bofors when in use: today, at 620 rpm, the noise is virtually continuous. Stinger is a hand-held rocket-propelled air defence missile, which launches a fragmentation warhead over a range of 3 miles at about 1200 mph. The latest version of Stinger has a complicated guidance system, which is said to give great accuracy.

The Soviets have produced several types of SAM, one of the earlier of the recent ones being SAM 7 (Grail), which has a fragmentation warhead and a range of 6 miles. A useful, easily handled weapon against small aircraft, it has been exported to many countries and is particularly popular with guerrillas and terrorists, who find it very useful when aircraft come to look for them. SAM 8 (Gecko) is much larger. The system, mounted on a six-wheel amphibious vehicle, carries six missiles and has a complex guidance system which includes an optical tracker. SAM 10, which has not yet acquired a nickname, is even more powerful: SAM 10s are thought to be included in the Soviet anti-ballistic missile system and therefore data about their performance is difficult to obtain. There seems little doubt that SAM 12 – or SA-X-12, to give it a more textbook title – is a powerful, dual-purpose interceptor, mainly designed to counter aircraft but also capable of intercepting and destroying short-range missiles. SAM 13 is a self-propelled air defence system which has a heavier-than-usual missile at 66 lb. Finally (for the moment) the Soviets have a self-propelled gun system in the ZSU-23-4, with a crew of four. Its four guns, each of which has its own guidance system, can dispose of 4000 rpm. It seems well capable of dissuading interested aircraft from investigating the sites it may be guarding.

Although the names of certain land-based surface-to-air missiles have become well known to the public in recent years, there is a lesser-known group of sea-launched surface-to-air missiles which are no less deadly. Britain has Seacat, an optically guided missile which launches a 140-lb HE warhead to a distance of 4 miles. Although primarily for use against aircraft, it also has a role against shipping. Many NATO ships are equipped with Seasparrow, which is based on a US prototype. Seasparrow has a range of 4 miles and delivers a 44-lb HE warhead. Standard is another American design, but has a longer range (17 miles) and a much larger HE warhead of 1280 lb. Standard is already widely distributed among the armies of NATO and is likely to be produced in large quantities for the foreseeable future. The British Seawolf is a shorter-range (4 miles) weapon which launches 180 lb of HE against oncoming missiles or aircraft. The Soviets have a corresponding range of missiles in the SAN range which has close links with the Gecko, mentioned earlier. The latest in the range is the SAN-7, which is reported to have a range of 17 miles, but few reliable details are known. The Soviets also have a range of surface-to-surface short-range missiles in the SS.N.9–19 range. SAN-12 (Sandbox) is thought to have a range of approximately 300 miles. These weapons are a form of cruise missile.

Air-to-air, and air-to-surface, missiles have been developed by all countries in large quantities. Most have elaborate guidance systems designed to make them home in on their targets while avoiding counter-missiles. As with all highly technical weapons, their complex nature gives them a certain unpredictability which is likely to be increased by developments in electronic warfare. Submariners who used the earlier torpedoes occasionally had the unnerving experience of realizing that their own torpedoes had missed the target and were sweeping round in a large arc that brought them perilously close to their launch point. No doubt it will be even easier to contrive a 'return to sender' device for airborne missiles, though as yet no one has ever hinted of one.

Two of the best-known air-to-air missiles, possibly because of their nicknames, are Sidewinder and Sparrow, Sidewinder (AIM-9) is the name given to a species of rattlesnake, but the missile bears little resemblance to one. It has a range of up to 12 miles and delivers a fragmentation warhead weighing approximately 175 lb. Sidewinders have been in existence for nearly twenty years – though, of course, being constantly improved. In the Falklands War Britain, which at that time had no Sidewinders, acquired 100 from America and found them effective. (Everyone is interested to see how a weapon performs under genuine combat conditions; it can be very different from results logged in practice sessions.) The latest version of the Sidewinder appears to be the 9 N.

Sparrow is an even more formidable weapon with a speed of approximately 3000 mph, a range averaging 50 miles and an 88-lb warhead which spreads an arc of lethal fragments on impact. Britain produced a similar weapon with a lesser range, entitled Skyflash. It was brilliantly successful, but production was abandoned some six years ago as it was cheaper to buy a similar US missile, the AMRAAM (advanced medium-range air-to-air missile).

The Soviets have also been active in this field, although they have been reluctant to disclose much information. It is, however, known that the Aphid is a short-range (approximately 3 miles) missile with a speed of about 2000 mph and a warhead of 12 lb. Aphid is properly AA-8 and has been followed by AA-9, AA-X-10, AA-XP-1 and AA-XP-2. All these are probably part of the Soviet anti-cruise system.

Not to be overlooked are the efficient Italian Aspide and the Israeli Python 3. Python has often been in action, mainly around Lebanon. It can deliver a 24-lb warhead over a range of 9 miles.

Air-to-surface missiles are designed to demolish much heavier targets than can air-to-air ones, and have a variety of warheads which can include nuclear. Britain has produced a formidable weapon in ALARM, a heavily computerized missile which locates its target in the most strongly defended area. Once it is programmed to its intended destination it is likely to reach it. Britain also has Sea Eagle and Sea Skua, both of which are specifically designed to deal with ships.

Among the French range of air-to-surface missiles the Exocet has gained a considerable reputation from its successes in Argentine hands in the Falklands, where it appears to have caught British forces by surprise. Exocets, specifically designed for use against ships, have a range of up to 40 miles and a speed of 750 mph, and carry a 364-lb warhead specially designed to penetrate armour. The later version of this weapon was the Super Etendard. The Falklands War was an excellent advertisement for the French armaments industry, for the Argentine-launched Exocets inflicted heavy losses on British shipping and Milans were used to great effect by the British infantry.

Both the United States and the USSR have an impressive array of ALMs (air-launched missiles). The USA has HARM (AGM-88A), which carries a fragmentation warhead at approximately 1500 mph over an 11-mile range. Their Harpoon is larger and, as its name implies, designed to penetrate armour; Maverick is smaller but probably faster than Harpoon, though with similar penetrative powers; Paveway is a free-fall, laser-guided missile which can be carried by a wide variety of different aircraft; Walleye is an 800-lb bomb which approaches its destination thanks to a near-perfect guidance system; and Shrike homes in on to enemy radar with a 145-lb fragmentation bomb. The Soviets have Kitchen, which delivers medium or large nuclear bombs, Kelt, Kerry and Kingfish; but these are less advanced than the AS-8-12 range mentioned earlier.

One might wonder how anyone can stay alive with such a variety of destructive weapons on land, at sea and in the air. But men have been saying that since the crossbow was invented. Many modern weapons are mutually cancelling, and apart from that comforting fact there must always be the thought that the world is now confronted with a 'no win' situation except for local, limited, conventional-type wars. The endless development of armaments worldwide is not so much due to the fact that they are needed, or even intended for ultimate use, but more because if one country has them, or might have them, others feel they must follow suit. Add to this the rivalry between the services within countries – on account of which the navy, army and air force jealously watch what their colleagues are getting and demand the same or better – and it is hardly surprising that the world has nearly become 'insanity fair', although it is probably some way short of reaching that point.

The 'ultimate deterrents' are, of course, the intercontinental ballistic missiles. Some of these are air-launched cruise missiles (ALCMs), others submarine-launched ballistic missiles (SLBMs). Some may be capable of being launched from air, sea or ground. The American Minuteman has a range of over 8000 miles and can deliver three nuclear warheads. It is launched from heavily fortified silos, but some of these are now being handed over to Minuteman's successor, Peacekeeper. Peacekeeper has the slightly lesser range of 7000 miles, but can launch ten warheads, each with the explosive power of 33,000 tons of TNT. There is some doubt about the wisdom of fixed sites for missiles, however 'hardened' their hiding places are, but the alternatives of mounting ICBMs on railways, or making them air-mobile, have not so far found much support.

The French, as is known, have an independent deterrent, the SSBS (sol-sol ballistique stratégique). This has a shorter range (2000 miles), which is adequate for potential needs; the SSBS has a thermonuclear warhead. The Soviet range of ICBMs includes the SS-17, SS-18, SS-19 and SS-20, all described in detail earlier in this chapter.

Formidable though land- or air-launched ICBMs are, there is greater apprehension about the less detectable presence of submarine-launched ballistic missiles (SLBMs). The French have a formidable contender in this field with the MSBS (met sol ballistique stratégique) M-4, which has a range of 2300 miles and carries six warheads, giving a combined explosive total of just under a million tons of TNT. The US contribution is Poseidon and Trident. Poseidon can carry ten warheads 3000 miles, or fourteen for 2500. Although Poseidon has now been surpassed in performance by Trident, it is still considered to be viable until the early 1990s. Trident has a range of approximately 4500 miles and carries eight 100,000-ton warheads. The latest version, the D-5, which has been ordered by Britain, has an even greater, though undisclosed, potential. The USSR has the SS-N-18 and the SS-N-20: the former has a range of 4000 miles and carries three 200,000-ton warheads, and the latter a range of over 5000 miles over which to deliver its nine warheads of much the same capacity. It must be remembered that all these figures are approximate: figures given for military potential are seldom strictly accurate, for obvious reasons.

9

The Chemical Arsenal

Bacteriological and chemical weapons, also denizens of the modern chamber of military horrors, are nothing new in principle. Chapter 1 described how poisoned arrows were made and used; later the poisoning of water supplies was discussed, and the medieval period is full of stories of dead men and horses, usually in an advanced state of decomposition, being catapulted into castles and fortresses. The main problem with such weapons is that their effects might work back to the user; anyone who poisoned a well, usually by filling it with rotting carcases, might easily have cause to regret doing so.

After the Middle Ages it was generally assumed that the use of such methods was unbecoming to civilized nations; it was fair and reasonable to chop someone to pieces, starve him or crush him under a massive bombardment, but to poison him was thought to be barbaric. This view did not occur to the Germans, who first used poison gas as a weapon of war in early 1915 in Belgium. The Allies had been told of its existence by captured German soldiers, but refused to believe what they were told. The Germans, who had kept their gas cylinders in the front line for several weeks before deciding to use them, were so surprised at the enormous effect on the French Algerian troops opposite that they failed to follow up their success. As the First World War progressed, the Germans quickly regretted that they had ever resorted to this particular form of firepower: the British used it more extensively and effectively – though not, as it happened, as much as they might have done. It was also an unpredictable weapon: when the British used gas in the autumn of 1915 the wind changed and blew some of it back on to the British lines. Protection was rudimentary at that stage, and men were advised that the best method was to urinate on to a piece of cloth which should then be tied over the mouth and nose. Gas masks, or respirators as the army preferred to call them, gradually improved during the course of the war, but so too did the range of gases, mustard gas and phosgene being the most effective and widely used. Everyone recognized that there was something especially vile about gas. It killed men and animals in a particularly painful and distressing way, and it also poisoned the ground, giving mud and water a loathsome sheen and smell. Nevertheless no one would have dreamt of foregoing its use.

Between 1918 and 1939 fear of future gas attacks was stimulated by somewhat extravagant scientific predictions; it was said that one small gas bomb would be able to kill the entire population of a city such as Southampton. The most likely gases were still mustard and phosgene, but it was said that the Germans (and probably the British) possessed a new one called arsine. In the event, gas was only used in one small attack by the Germans in Poland, but throughout that period both sides maintained large stocks. Often these were kept very close to the front line and caused considerable anxiety. On several occasions the gas storage containers looked like being caught up in the general hostilities, and perhaps activated by enemy shellfire (see Chapter 12).

In 1916 Britain established a 3000-acre experimental gas warfare centre at Porton Down on Salibury Plain; soon it was one of the most successful enterprises in the world for this type of research. Much of the gas used at this stage was phosgene, which had so little smell that it was often inhaled in lethal quantities before the unfortunate victim realized what was happening. However, mustard gas had considerable advantages as a battlefield weapon. It was not quite as unobtrusive as phosgene, but it was thought to be more dangerous as it attacked the eyes and could also burn the skin through clothing. The casualties of gas warfare were by no means limited to the battlefield. Before the importance of taking proper precautions was understood, factory workers engaged in its manufacture suffered from various forms of burns. The result was that many of them lost their voices temporarily and some suffered permanent damage; eye trouble, particularly conjunctivitis, was widespread, and their skin began to itch and peel.

Although gas had been an effective battlefield weapon in the First

World War, it was heartily disliked even by those who benefited from its use. Gas cyclinders were awkward and heavy to transport and their mere presence produced a kind of revulsion which was not extended to high explosive. Nevertheless there was a considerable range to choose from by the time the Second World War arrived. Many of the gases available had, of course, been tested twenty years earlier and their properties were well known. The Germans, after being the first to use gas, had eventually suffered from it more than the Allies; therefore in 1939 they ensured that they had full stocks and the means of delivery. Gases were classified as choking, such as phosgene and chloropicrin; poisonous, such as hydrogen cyanide; blister, such as mustard and nitrogen-mustard; and harassing, such as Adamsite and Lewisite. Gas warfare had become unpleasantly subtle. A preliminary gas, such as chloroacetophenone, could cause the recipient to sneeze and to remove his respirator for greater comfort; thus exposed, he would be an easy victim for something more lethal, such as phosgene. (Tear gas is not, of course, lethal but can produce temporary blindness and violent choking.) It is worth noting that nerve gases, often thought of as being the latest of the products of this diabolical Pandora's box, were being produced by the Germans as early as 1942 (tabun) and that an even more efficient version (sarin) was due to go into production in 1945.

Gas, like other forms of chemical warfare, is a particularly intimidating weapon when employed against civilians. In the Second World War many German citizens had no form of protection against gas attack. Every civilian in Britain, plus certain animals whose owners could afford special equipment, had a form of gas mask and was supposed to carry it everywhere he or she went. The fact that the civilian gas mask would not have protected its owner from the sort of gases which the Germans might have used was not considered so important as ensuring that everyone *thought* he or she was protected, and therefore would not panic at the warning of a gas attack. Ironically, in the 1980s the public was warned that asbestos had been used in the make-up of gas masks and that they were now regarded as lethal in their own right.

The wartime government seems to have been somewhat naïve in assuming that every civilian would carry that little square box wherever he or she went. Most people lost them regularly: cinema managers collected them by the thousand, an experience shared by workers on trains and buses and in restaurants. Soldiers and other government employees were ordered to carry a respirator, which was of a superior type to that issued to civilians. Its principal weakness was that it was held in a haversack which gave no hint of the contents. Soldiers going outside camp were therefore prone to remove the respirator and replace it with a pair of shoes. They would then change out of their heavy army boots and endeavour to cut a dashing figure in the local dance hall.

Gas is not the most cost-effective method of chemical warfare, and even in the Second World War was only one of a range of options, though the effects of these could be even less predictable than that of gas. Bacteriological warfare was even more horrific. In 1942, when Britain learned that Germany and the Soviet Union were both experimenting with germ warfare, British scientists developed an anthrax bomb and tested it on Gruinard Island, off the coast of Scotland. Gruinard was considered to be so dangerous after these experiments that it has remained forbidden territory for over fifty years. In 1987 it was pronounced sufficiently clear for a flock of sheep to be allowed to graze there. If fifty years is a reasonable sample of the time required for land to become 'clear' after the use of bacteriological bombs, this in itself might well help to prevent their use in any future conflict: few belligerents are prepared to acquire land for the occupation of the third generation after their own. Nevertheless, by 1944 the British government was ready to saturate German territory with several hundred thousand anthrax bombs if the Germans, known to hold large stocks, were so unwise as to begin using them. And this was only one of the retaliatory options Britain had in hand. There were times when the Second World War really looked like being the war to end wars – and the human race in the process.

Instruments of chemical, biological and bacteriological warfare differ considerably in the size of their effects and the means by which these are produced. One of the slower-acting ones is the herbicide, which, sprayed on enemy crops, will reduce their owners to starvation; the means, rather than the method, seems the only difference between this and its use in forms of warfare mentioned earlier in the book. It is, of course, almost impossible to protect crops from attack; protective covering, such as plastic sheeting, can easily be burnt away. Reports of the use of this type of weapon have come in from many different sources in the last few decades. The United States used herbicide in the war in Vietnam, but stopped short of the range of chemical agents which the Vietnamese subsequently used against their neighbours. This was described by the victims as 'yellow rain', although it could be white, red or green. Those who suffered the heaviest doses died from vomiting blood; others, less severely affected, found their skin itching and blistered; diarrhoea was widespread.

Similar stories emerged from Afghanistan, where spraying from helicopters was the main method used. Spraying from aircraft is, of course, a common means of pest control, and is extremely effective in improving food supplies when used constructively. Even before yellow rain was used in Afghanistan, it was thought to be of Soviet origin; the fungus from which it is derived grows in the USSR and in very few other places. As with all chemical/biological weapons, the first effects observed may not be the last. The misuse of certain deadly chemicals is likely to upset the balance of nature in more senses than one, as those who imagine they have hardly been affected by yellow rain may come to find out.

Agent Orange, which contains dioxin, was used by US forces in Vietnam as a jungle defoliant; its purpose was to remove cover from the Viet Cong forces and expose their hide-outs, rather than to kill them by its use. The effect on the Viet Cong has not been scientifically assessed, but some US soldiers who suffered mild exposure to Agent Orange subsequently reported physical damage to themselves and genetic damage to their children. Some of the effects of lethal gases were the subject of studies made by the Germans in their concentration camps, and also by the Japanese, who experimented on British prisoners of war in the Far East; it is difficult to see how nations which condone such activities can think themselves civilized.

Attempts to prevent the stocking of various forms of toxic agents have not been as successful as was hoped by those who agreed to the biological warfare convention of 1972. The United States destroyed vast quantities of wartime stocks and apparently did not replace them; however, a few years ago it was announced that the US lead in this matter had not been followed by the Soviet Union, which held enormous stocks and was training troops in their battlefield use. Somewhat belatedly, the United States is trying to recover the lost years of research and manufacture.

One use of toxins as a weapon came to public notice in 1978 when a Bulgarian defector named Georgi Markov died of unknown causes in

London. He was thought to have been assassinated while waiting at a bus-stop. The cause of his death was found to be a minute pellet containing ricin, an immensely powerful poison extracted from the castor bean. Ironically, it has been studied as a potential cure for cancer, for when it enters the bloodstream it prefers to attack diseased cells in preference to healthy ones – though eventually it attacks the healthy ones too. There is no known antidote, although ricin has been in existence for at least fifty years; it is known to the US forces as WA. Markov's pellet may have been inserted by a prod from the point of an umbrella, which was hardly noticed at the time. This killing opens up the whole field of political and military assassination, as ricin and similar toxins are very difficult to detect. Markov was not of any great military or political value, although he had become hated and feared by the Bulgarian government, but the possibilities of this type of assassination are alarming, to say the least. From time to time the premature death of the British Labour Party leader Hugh Gaitskell in 1963 has been regarded with suspicion. Gaitskell espoused several unpopular causes, and his removal from the scene of influential politics would have been viewed with satisfaction in some quarters.

It is impractical to ban such weapons, because breaches of the ban would be virtually undetectable. Furthermore – as may yet be possible with ricin – many deadly poisons have positive, healing effects if used in the correct way. In nineteenth-century chemists' shops nearly all the drugs and disinfectants were labelled 'Poison' – as indeed they were. Strychnine, for instance, which can produce an agonizing death, is also the means of prolonging the lives of heart sufferers. One can but hope that these deadly minutiae of firepower will be used only for therapeutic purposes in future.

Two aspects of chemical/biological warfare may make these weapons appear somewhat less appalling. The first is that the spread of disease and its abatement seem to depend very often on natural and scarcely understood causes which may limit their effect. Bubonic plague is endemic in certain parts of Asia, but, like cholera, tends to become dormant. When there is an outbreak it spreads rapidly but, even without human preventative methods, may suddenly abate. There are various explanations for this, one being that outbreaks usually occur during the rainy season, when rats are driven out of their normal habitat and water supplies are contaminated. But such explanations are usually incomplete. The British Army in the Crimea in 1854 was devastated by cholera, even on the journey out, but mysteriously the epidemic suddenly seemed to have run its course. Possibly an immunity builds up. The history of the spread of syphilis suggests that once it is firmly established in a community its effects tend to lessen. Similarly, the common cold can prostrate people who are infected with it for the first time; yet the average British citizen finds it little more than an inconvenience.

The second advantage of using gas rather than other weapons in warfare is that certain gases can be used to paralyse but not kill. Such gases are often known as hallucinogens, for they can produce disordered thought, and even muscular paralysis, without killing. Nearly all these toxins, such as ergot, lysergic acid, opium and psylocybine, have been known for centuries and are used, in various formulae, for producing trance-like states in those unwise enough to experiment with them. However, properly disseminated, perhaps by means of bombs or sprays, they could induce whole armies to sit down quietly while being disarmed.

The greatest threat from ultra-sophisticated, non-explosive weapons is the use which could be made of them by terrorists. Superpowers would probably consider them less effective than other weapons at their disposal, but not so the guerrilla or terrorist. There is, unfortunately, evidence that small groups have already been experimenting with nerve gases and botulin – the latter for the contamination of food supplies. The methods used by terrorists involve holding a person or a nation to ransom. Contamination of buildings through the ventilation systems, or of water through reservoirs, are the easier option, provided the necessary materials could be acquired and the 'bomb' manufactured. In the event, though, this form of firepower might be counter-productive. Apart from exposing the terrorists themselves to the same, or even greater, dangers, the effect would be to alienate all possible sympathy and support for their cause. If, therefore, a threat of this nature ever developed, it would probably be due to the acts of a 'mad scientist' of the variety much favoured by Victorian fiction writers and more recently in Ian Fleming's James Bond novels. Such a person could cause considerable trouble before the results of his efforts were neutralized but, fortunately, such characters have not yet appeared in real life. It is to be hoped that they never step out of the pages of fiction and create their Frankensteins in the field of germ warfare.

10

The Role of the Navies

As the aim of the owner of any form of firepower is victory, it is obvious that the chances of achieving it are all the greater if he can bring that firepower to bear at the earliest possible moment. This point has been made many times, though seldom as pungently as by the American poet Henry Wheeler Shaw in 1865:

> Thrice is he armed that hath his quarrel just
> But four times he who gets his blow in first.

There is truth in that saying, but not the whole truth. Germany got her blow in first in Poland in 1939, and in France and Norway in 1940, but eventually was ground to defeat under a superior weight of metal. Japan struck the United States and Britain without warning at Pearl Harbor and in Malaya in 1941, but eventually her armies and navies were rolled backwards and she received the *coup de grâce* of the first atomic bombs. Both defeated countries neglected the cardinal military maxim that force should be concentrated until final victory is achieved. Germany eventually found herself fighting on fronts in France, Italy and the Soviet Union; Japan had her armies dispersed over a wide range of Pacific islands and was unable to support them adequately by sea or air. At the same time she was outfought in Burma and frustrated in China.

Yet for centuries it has been clear that correct strategy combined with an adequate delivery system is an essential ingredient of victory. When Napoleon marched towards Moscow, his supplies proved to be inadequate from the moment he set out, yet he should already have learnt his lesson from fighting in the peninsula – in Spain and Portugal where, as Wellington put it, small armies are beaten and large armies starve.

In the present century military planners have recognized that the means of delivery of weapons are almost as important as the weapons themselves. In consequence some weapons incorporate their own delivery systems; the self-propelled gun is an example. In the Western Desert, when the Special Air Service planned to eliminate German aircraft on the ground they decided to arrive by parachute. Soon, however, they found that coming by vehicle or on foot could be less conspicuous. They continued to train as parachutists, but regarded the parachute as only one of several alternative methods of travel.

Little is known about the land supply systems of early armies, though much information has become available about their sea transport. Armies have made great use of horses, although the horses' fodder requirements created almost as many problems as their pulling power solved. Elephants have been used for centuries in the Far East, but when Hannibal used them in Europe his aim was to terrify the enemy rather than to solve his logistical problems. In the nineteenth century, General Garnet Wolseley used portable canoes to enable his troops to reach the remote stronghold of a rebel leader in Canada, and arrived so rapidly that the man departed leaving his breakfast on the table. General Kitchener built a railway to enable him to overcome the Dervishes and win back the Sudan. In the First World War, when troops from England had to cross the English Channel to fight in France, and then come back on leave or to recover from wounds, twenty million men crossed the water between Dover and Calais and Folkestone and Boulogne. Had Germany been able to cut that lifeline with submarine and surface activity, the war might have ended very differently, but thanks to the Royal Navy it did not.

In the Second World War aircraft were nothing like as advanced as they are today and the maximum load which could be carried in a B29 was 10 tons – the capacity of one railway wagon. Railways and ships were therefore the most important means of transport. Ships carried huge cargoes of men and munitions across the Atlantic, protected, as far as possible, in naval convoys. Railways, and to a lesser extent lorries, conveyed men and supplies overland. When the Allies were advancing across Europe after D-Day to strike at Germany, it became clear just how much firepower depended on sea and land transport.

Until the Channel ports were captured, supplies had to come in from Cherbourg or over the artificial harbour at Arromanches. Although some supplies were offloaded directly on to the shore, they were but a small fraction of the total requirement.

Unfortunately, the successful use of road transport in the later stages of the Second World War created the impression that one could not have too much of it. This was found to be untrue when the next war occurred in Korea, a country with a very limited road system. Within a short time all roads were hopelessly clogged with trucks and other vehicles and it became necessary to have a strict system of regulations. Meanwhile, the Chinese made excellent progress on their feet. The wheel, so to speak, had turned full circle.

In the past there has been much confusion over the use to which sea transport should be put. With some exceptions the general attitude towards transports was that the crew who sailed them should be distinct from those whom they transported. In 1588 this policy proved the undoing of the Spanish Armada, whose vast ships were made less seaworthy by an enormous superstructure like a medieval castle (hence the forecastle and aftcastle), which resembled a floating barracks. These ships had firepower which they could not bring to bear on the smaller, manoeuvrable ships under Drake's command. However in all navies, including the British, there was always a tendency to load a ship so heavily with guns that it lost much of its speed and manageability. Nelson's ships were successful at the Battle of Trafalgar in 1805 because he was a superb tactician and additionally had ships whose large crews could turn their hands to almost anything.

In the twentieth century the need to move much larger quantities of men and materials has once more widened the distinction between fighting ships and transports. It was soon discovered that no normal passenger ship or freighter could protect itself against the U-Boat, which in the First World War could approach almost unheard and unseen. Protection had to be provided by the aptly named destroyer and other fast, heavily gunned craft. Subsequently a submarine had to approach its prey very swiftly and fire its torpedoes accurately if it wished to succeed. Having done so it needed to make its escape quickly, or a destroyer would be after it with depth charges, which exploded with such force under water that the submarine's structure would be damaged by the shock waves.

By the time the Second World War broke out, naval warfare had become very sophisticated. Mines, which had been used to considerable effect in earlier wars, could now be laid by aircraft rather than having to rely on a submarine or surface craft. But, equally, the sea could be surveyed by aircraft. It was obvious, though the navies of the world were reluctant to admit it, that command of the sea no longer rested with the battleship or its smaller sisters, but with submarines and aircraft. In pre-war exercises anti-aircraft guns on battleships had seemed equal to the worst that attacking aircraft could do; unfortunately, when the *Prince of Wales* and *Repulse*, completely lacking in air protection, were caught by waves of Japanese torpedo-bombers off the coast of Malaya in 1941 they did not last long.

But soon it was the Japanese turn to experience the potential of air power at sea. The Japanese had inflicted a damaging blow to the American fleet at Pearl Harbor in Hawaii in December 1941 and had therefore assumed that the USA would be helpless to retaliate in the Pacific. But they had reckoned without American resource and industrial capacity. The United States began building not merely aircraft carriers but also the means to repair them, and other shipping, while they were still at sea. When Henry Kaiser's shipyard was given an order to build an aircraft carrier, he took eighteen months to build the first. But the month after the first went down the slipway, another was launched. Others then followed at the rate of one a month.

The early sea battles were won not merely by aircraft but by another aspect of firepower – the ability to know what the enemy plans to do next. War, in the future, was going to be the war of code-breaking. Spying and confidential codes had always been an essential part of warfare, but never before had they been so important as in this one. Through reading the Japanese codes, the American naval commanders were able to outmanoeuvre their opponents. The part played by code-breaking in the enhancement of firepower will be discussed later.

But the role of the heavy battleship was not yet over, even though at the decisive Battle of the Coral Sea in 1942 the American and Japanese fleets were never in sight of each other. The battleship could still help to preserve the carrier from surface attack, particularly when the latter's aircraft were engaged elsewhere. Radar, and the proximity fuse which enabled the battleship to intercept missiles, did much to prolong its life. Better armour and a wider variety of guns made it less vulnerable. As a result, battleships survived a variety of attacks, including those by torpedo and kamikaze aircraft. Kamikaze aircraft, usually small planes built on the lines of an aerial torpedo, were flown to their destination by pilots who crashed them on to the target, if not shot down before arrival. (Their naval equivalent was the kamikaze submarine, which often carried a crew of several men.) Kamikaze aircraft were but a short step from the dive-bomber which, though not intended to be a suicide plane, would not always pull out of its dive in time. On occasion the Japanese used kamikaze tactics with larger aircraft.

For the Pacific theatre the United States devised a series of balanced fleets known as task forces, which contained a mixture of battleships, cruisers, carriers and destroyers and in some areas also included smaller craft. Naval warfare became very sophisticated in the Second World War: it was no longer a matter of huge ships sending enormous shells crashing towards each other, or even of destroyers versus submarines, but extended to small underwater craft which would approach larger craft secretly in the darkness and attach limpet mines. A battleship in harbour might now be more vulnerable than one at sea, from either air attack or midget submarines, as the huge German battleship *Tirpitz* found to its cost in 1944.

Since 1945 all the major powers have possessed aircraft carriers. France has the *Clemenceau* and the *Foch*, both of which have a crew of over 1300 and a 32,000-ton displacement. Each carries twenty-eight aircraft, of which sixteen are Super Etendards. In addition to the aircraft an unspecified number of helicopters are carried. These carriers make a formidable pair, but will look antique when they are replaced by nuclear-powered carriers in the next decade. Italy has a light carrier of 13,000 tons which, with eighteen Sea-King helicopters and a variety of sophisticated weapons, is mainly designed for anti-submarine work.

Britain's four light carriers (19,000 tons) are equipped with Sea-Harrier strike aircraft and Sea-King helicopters. *Hermes*, a somewhat elderly craft, and *Invincible* performed extremely well in the Falklands. *Ark Royal* and *Illustrious* are newer (1982) and incorporate a very useful range of weaponry. Britain has been a pioneer in carrier development and invented both the angled deck and the steam catapult for assisting landing and take-off.

The USSR has four substantial carriers, each of 42,000 tons, in the Kiev class. Although these carry thirty aircraft, they are primarily for anti-submarine warfare and include two squadrons of helicopters. They are also equipped with eight SS-N-12 launchers.

The United States has four carriers in the Kitty Hawk class and four in the Nimitz class. The Kitty Hawks, which displace 80,000 tons when fully loaded, carry a crew of over 5000, seventy aircraft and sixteen helicopters. Nimitz carriers are even larger, and displace 91,000 tons when fully loaded; they are nuclear-propelled. Armament and aircraft vary slightly among the different ships, but an average figure is sixty-five aircraft plus sixteen helicopters. Two more ships in the Nimitz class are on order in 1987.

The Soviet Union and the USA both maintain their faith in the value of battleships. Russia already has two battle cruisers in the Kirov class and two more under construction. There is, however, a considerable difference between the old-style battleship and the new one. The modern battleship relies more on missile launchers than guns, although she also carries the latter. The Kirovs are in many ways a mobile rocket platform, for each has provision for twenty SS-19-launchers, twelve SA-N-6s, two SA-N-4s, two SS-N-14s and two RBUs. Propulsion seems to be by a combination of nuclear and steam power.

The US Iowa class is considerably larger, displacing 38,000 tons at full load. In one sense these are modernized museum pieces, for they all date back to 1943. Their armour consists of a mixture of missile launchers and guns, and in spite of their age they are very formidable craft.

Although the cruiser (faster, but smaller and more lightly armoured, than a battleship) was thought to be obsolete in the 1970s, several countries possess them and are continuing to build them. The Soviets have two in the Slava class, displacing 12,000 tons, and two more apparently under construction. Slavas carry an array of rockets and are propelled by gas turbines. Udaloys are much smaller, displacing 8000 tons; they are similarly propelled, but carry a substantial quantity of rocket launchers. The United States has the Virginia and the Ticonderoga class. The former is nuclear-propelled and displaces 11,000 tons. It carries two helicopters, a crew of 360 and a variety of rocket launchers. The Ticonderogas (two in hand but eleven more in prospect) displace 9000 tons, are gas turbine-propelled, and have the usual complement of launchers and torpedo tubes.

The next step down brings one to destroyers and frigates. A normal displacement figure for a destroyer should be around 5000 tons, but the Soviet Sovremenny class at 8000 tons, and the US Spruance/Kidd, at approximately the same, both exceed this. The Sovremenny class carry the SS-N-22, as well as a variety of other missiles. The Soviets seem to have fewer of these ships than one might have expected. The US Spruance/Kidds have a very sophisticated armoury and guidance system and are gas turbine-propelled. France has a very useful small destroyer in the Georges Leygues (C.70) class, although these ships displace a mere 4000 tons. They are very efficiently equipped for anti-air and anti-submarine warfare.

The British Broadsword class, of 4000 tons, proved itself in the Falklands where it also learned some useful lessons about the necessary balance of armaments. The Broadsword class carry Exocet and Seawolf missiles. Six of these ships are available, four are under construction and three more are on order.

China has a surprisingly large number of 4000-ton destroyers, all fourteen of which are equipped with a variety of weapons for anti-submarine and anti-air warfare. The Chinese intend to increase the number of these vessels. The Japanese sea-defence component of their 'self-defence forces' (the name given to the limited number of armaments permitted by the agreement of July 1954) has two 7000-ton destroyers, and two more on order. Their primary role is anti-submarine warfare and they are well equipped for it.

Frigates – which are small, averaging 3000 tons – are thought to be one of the most useful ships for any future naval warfare, and so most countries have a good complement. The USSR has thirty-two in the Krivak class. These displace under 4000 tons and carry a mixture of armaments which include four SS-N-14 launchers and two twin 76-mm guns. The United States has thirty-six O. H. Perry frigates of approximately 3600 tons displacement, which carry two helicopters and include the M13 Harpoon launcher in their armament. Britain has the smaller, 3250-ton Amazon class, which carries four Exocet launchers as well as a Seacat launcher.

There is also a range of small, versatile craft, usually under 5000 tons, mainly diesel-propelled and frequently hovercraft. Often the latter are known as ACLCs (air cushion landing craft). They have a variety of uses from patrolling to minesweeping and could make a useful contribution to landing troops, though this function would more easily be discharged by much larger craft capable of handling tanks. The Soviet Rogovs displace 11,000 tons, but the US Tarawas extend to 39,000. The United States gained considerable experience in the use of landing craft in the Pacific during the Second World War; having been driven from the Philippines to Australia, she fought her way back to Japan by island landings. Names like Iwo Jima and Okinawa will live long in American history on account of the bloody battles fought on their beaches.

The firepower which even quite small vessels can bring to bear is enormous. Anti-submarine weapons include the British Stingray, an acoustic homing torpedo which can be launched from surface ships, helicopters or fixed-wing aircraft; the US Navy's ALWT (advanced lightweight torpedo) Asroc, which is a form of nuclear depth charge; ASW-SOW (anti-submarine stand-off weapon), and Subroc. Subroc is a torpedo with a nuclear warhead which rises to the surface, travels by air and descends. The USA also has Captor, a moored torpedo which can be released at the appropriate moment. Australia has a useful weapon in the Ikara, which carries a torpedo by missile to the target area and then allows it to parachute down. This method means that the torpedo takes less time to arrive in the target area than if it had made the first part of its journey under water. Soviet surface ships all carry a variety of rocket launchers, and the Silex (SS-N-14) appears to be similar to the Ikara system. The SS-N-15/16 is launched from a submarine under water, rises to the surface to travel to its destination, and when in the target area releases either a nuclear depth bomb or a torpedo. The French, the Swedes and the Italians also have useful anti-submarine weapons, but none more powerful than the ones described above.

All countries also have an impressive range of naval guns. Italy has made considerable strides in this field: the OTO 76/62 COMPACT can fire eighty rounds a minute without reloading. The Soviet equivalents are the twin 76-mm, which fires sixty rounds a minute from its twin-barrelled system, and the 100-mm, of which the official rate of fire is not available. Invariably, the larger the gun, the slower the rate of fire. The USA has some enormous guns in the triple 16-inch class which only fire two rounds a minute, but if one shell hit the target a second would not be necessary. Slightly lighter are the American 5-

inch guns which can fire up to twenty rounds a minute. Both the USSR and USA have designed rapid-firing (3000 rpm) guns for shooting down oncoming missiles. Both types have six barrels, but the Soviet version has a slightly larger calibre – 30-mm against the US 20-mm; however, the latter seems to have an exceptionally efficient guidance system and has been adopted by Britain, Australia and Japan.

In the 1980s the submarine has acquired the status of the most important card in the 'war game'. (War games are nothing like ordinary games: they are exercises, often conducted during conferences, when the relative options of potential opponents are carefully assessed.) Ever since the First World War, when Germany inflicted enormous damage on Britain by unrestricted U-Boat warfare, the submarine has been steadily developing to the point at which it is now one of the most essential weapons in a nation's armoury. An example of its power was shown in 1944 when a US submarine, the *Archerfish*, sank the latest and largest Japanese aircraft carrier, the 59,000-ton *Shinano*, before the latter had moved out of coastal waters or fired a single shot. US submarines sank seven other Japanese carriers during that war, as well as a host of smaller ships. Germany also had significant successes with submarines in the Second World War, so much so that in 1943 they came close to achieving final victory by this means.

Post-war developments soon showed that the submarine was not likely to be moved from its position of vital importance in military and naval conflict. In 1959 US Polaris submarines demonstrated their ability to carry and deliver a nuclear missile to a target over 1000 miles from its position; this gave warfare a new dimension. Nuclear power was already in use for propelling submarines; the USS *Nautilus* had been commissioned in 1954, and proved herself capable of travelling at 30 knots at about 1000 feet below the surface. The fact that a submarine could now travel vast distances under water and stay submerged for months on end – but could, if commanded to do so, emerge and send a missile into the heart of an enemy country – was a daunting prospect. It is a reasonable assumption that the superpowers now have hundreds of submarines, many of them nuclear, cruising around in the ocean depths, plotting and tracking each other's possible courses and intentions. A range of hunter-killer submarines also exists, as well as the aircraft and helicopters allocated to this particular activity. But it is not only the monster submarines which pose a threat: there is known to be a range of smaller submarines which can, no doubt, imitate and surpass the deeds of their forebears.

The intangibles of underwater warfare have given a vastly increased importance to all forms of electronic surveillance. Submarines are not difficult to detect if one knows where to look, but the sea is a large area. In undersea warfare it is important to know the capabilities as well as the possible intentions of potential opponents; some of this knowledge may be gained by electronic surveillance, but perhaps even more by spying. The NATO countries are, unfortunately, vulnerable to the activities of spies, who are allowed facilities they would not find in the Soviet Union. NATO and Soviet exercises are carefully monitored by the potential opposition, and NATO countries are only too well aware that all Soviet trawlers carry sophisticated electronic surveillance equipment. In 1987 Britain took part in a naval exercise in which commercial shipping was involved. Many of the merchant ships had Polish officers on board, a fact of which Soviet Naval Intelligence was well aware. As the ships were registered abroad and not controlled by unions, and as they operated as cargo-carrying ships more cheaply than the Merchant Navy, they were normally employed on British trade routes. The British Department of Transport should have made this clear to the Ministry of Defence before the latter sent the Polish officers secret details of the exercise, assessments of Soviet submarine movements and much other information greatly appreciated by the Warsaw Pact countries. But this lapse of security is as nothing compared with the extraordinary story revealed by the trial in the USA in 1986 of John Walker, a retired Navy communications specialist. Walker pleaded guilty to supplying naval secrets to the KGB for eighteen years. According to Secretary of Defense Caspar Weinberger in 1987, Walker had betrayed over a million vital secrets, a fact which had enabled the Soviets to improve their own naval capabilities dramatically while acquiring information about the United States defence services 'which would have been devastating in time of war'.

Not surprisingly, in view of the Walker revelations, the United States and the Soviet Union deploy many similar types of underwater craft. The USA has some thirty – perhaps fifty – 6000-ton nuclear-propelled submarines in the Los Angeles class. They carry Harpoon and Tomahawk missiles. The USSR has at least six – probably more – 3500-ton, nuclear-propelled submarines in the Alfa class, which compensate for their smaller size with their considerable speed. It also has an interesting submarine in what NATO describes as the Mitre class (8000 tons on the surface), which is nuclear-propelled and carries several SS-N-21 missiles. The hull of this type of submarine, of which six more are said to be under construction, is believed to be of titanium. This must be one of the most expensive submarines ever built, but presumably the Soviet Union must believe it is cost-effective.

The Soviets are making no effort to economize on submarine building. The Delta III class is a nuclear-powered, 11,000-ton submarine which carries sixteen SS-N-18 missiles with a range of nearly 5000 miles. The largest submarine ever built is the Typhoon class, with a 25,000-ton displacement and twenty SS-N-20 missiles. It is nuclear-propelled, but its speed is relatively slow. Meanwhile the United States has twelve sizeable submarines in the Ohio class, with at least another twelve on order. These displace nearly 19,000 tons, are nuclear-propelled and carry twenty-four Trident missiles each. Not to be outdone, France has produced its own nuclear-propelled, nuclear-armed submarine in the shape of six Redoutables, which displace just under 9000 tons each. Britain has four Resolution class submarines of 8400 tons, which carry sixteen Polaris Chevaline missiles; four more are on order. Britain has other nuclear submarines in the Trafalgar class (four plus two more on order), which displace only 5000 tons but have a useful turn of speed.

France has one of the smallest nuclear submarines in the Rubis class (six, plus four on order), which displace less than 3000 tons. They have a variety of armaments and, although their speed is given as some 20 knots, this figure may well be a considerable under-estimate. Russia also has eighteen 6000-ton Victor-IIIs, a nuclear-propelled class which carries both missiles and torpedoes; at least three Oscars of 14,000 tons, which carry twenty-four SS-N-19 missiles; a probable seven Kilos, of 3000-ton displacement; and twenty-one Tangos (3700 tons), which are conventional diesel-powered submarines. In this smaller conventional class, Britain has the Upholder: the number in service with the Royal Navy is likely to be ten. To this array of modern sea-serpents, West Germany contributes nearly sixty 1200-ton diesel-powered submarines, and has also supplied a number to other NATO countries. It also has some larger submarines in the TR 1700 class, diesel-powered craft which carry six 21-inch torpedo tubes. So far only two have been built and these have been sold to the Argentine Navy, a transaction which cannot have been welcomed by Britain.

Apart from all these later models there are undoubtedly many more conventional submarines in the navies of countries throughout the world. The massive piles of armaments stocked by all countries would be ludicrous if they were not so potentially dangerous, but nowhere is the proliferation of weaponry so bizarre as under the sea. The possession of a powerful navy has been a matter of national prestige for centuries. The USA and USSR are relative newcomers to this contest for the supremacy of the seas – the USSR because it felt humiliated by the Cuban missile crisis of 1962 and set to work to build a massive navy without regard to cost. Neither the Soviet Union nor the United States really needs a large submarine fleet, whereas Britain and France do. Nevertheless, fear of a 'sneak' attack or of one major power obtaining a decisive supremacy over the other will ensure that the oceans of the world will be patrolled by vastly expensive pieces of machinery for the foreseeable future: 'Those whom the gods wish to destroy they first make mad.'

A small sardine and its mother were swimming side by side in the sea when the huge, dark shape of a nuclear submarine moved overhead. 'Oh Mummy, what is that?' cried the small sardine.

'Nothing to worry about, dear,' said its mother. 'It's only a tin of men, they won't hurt you.'

Let us hope she was right.

11

Air Firepower

Man's use of the air for adding to his firepower began as long ago as the balloons of the late eighteenth century (see Chapter 6). In 1794 the French, who were fighting the Austrians, sent up an observation balloon with an intrepid captain as its occupant. Balloons developed into powerful airships, of which the German Zeppelin of 1908 was the most famous. Zeppelins had a speed of 40 mph and could carry five 100-lb and twenty $6\frac{1}{2}$-lb incendiary bombs.

It is said – though this fact is now disputed – that the American Wright brothers were the first to fly a heavier-than-air aircraft, in 1903. After that, progress along the road to Armageddon was rapid. A military aircraft was delivered to the US Army in 1909; it was first used for dropping bombs and as a gun platform in 1910. The following year saw an aircraft land and take off from the deck of a warship, as well as land and take off from water. For good measure, in that year an Italian launched a torpedo from an aircraft, and two Americans demonstrated that aircraft could be used to drop high explosives. A year later an American was proving that a Lewis (machine) gun could be used successfully in a similar machine, and the British tried the novel experiment of firing a small cannon shell against submarines. Anti-submarine warfare was very much in the minds of the early aircraft designers. Radio also made rapid progress: by 1914 it was possible for one pilot to stay in contact with another up to 10 miles away. All that was now needed was for aircraft to be given a thorough test in war. The Italians had tried out aircraft in their war against the Turks in 1911, but had concentrated on reconnaissance. The Turks were not long in acquiring their own aircraft from Germany, but these events were little more than skirmishes in relation to what was soon to occur in the First World War.

On 30 August 1914, soon after the outbreak of hostilities on 4 August, bombs and leaflets were dropped on Paris. On 5 October a German aircraft was shot down by machine gun fire from a French Voisin 89. Later in the same year, on Christmas Eve, German aircraft dropped a few small bombs on Dover. The pattern of aerial warfare had now been established. However, the heavier-than-air machines were noticeably less efficient at bombing than the Zeppelins were, as the east coast of Britain would soon bear witness. The worst raid was on London in September 1915, and was thought to be a revenge raid – the previous month, a Zeppelin had come down in the sea off Ostend following an attack by anti-aircraft guns at Dover. Zeppelins were more vulnerable by day, so naturally preferred to do their work by night: in order to frustrate them it became obligatory to turn off all lights if Zeppelins were thought to be heading for the area. This gave the English language a new word: the blackout. When the two world wars were receding into history 'blackout' came to mean any complete loss of consciousness, but no one who has experienced that method of frustrating bombers will ever forget the bizarre situation of unlit streets, completely covered windows, regulations enforced by air raid wardens who had the same popularity as modern traffic wardens, and searchlights crossing the sky as they tried to locate enemy aircraft. As shown earlier in this book, firepower is only as effective as the defence allows it to be; a bullet which cannot penetrate is no threat at all, and a bomber which cannot find its target has effectively failed, even though it drops its bombs on another part of the enemy territory.

Bombing was notably erratic until efficient sights were developed in 1915. The American firm of Sperry soon took a lead with their gyro-stabilized sight. Another considerable step forward was made when it was realized that machine gun bullets could be synchronized to fire their bullets through the rotating blades of the propeller. Inevitably, as aerial combats increased in number, certain pilots would show exceptional proficiency; most of them fully deserved their reputation as air aces, although few lived long enough to enjoy it. Among them were the German Max Immelmann and the Englishman Victor Ball, VC. Immelmann gave his name to a combination of diving and turning: the attacker would roar down on to his victim but, if his first spray of bullets failed to gain victory, would pull up into the vertical before turning and diving again. The Immelmann turn was extremely popular with aircraft which found another on their tails – a swift upward climb, a sharp turn, and the attacked could now become the attacker.

Certain aircraft soon began to gain a formidable reputation. One was the product of the Dutch designer Fokker; it had been rejected by Britain before it was snapped up by the Germans, causing much subsequent breast-beating. The British produced an excellent, highly manoeuvrable aircraft in the D.H.2 (De Havilland) and the sturdy Vickers FB5; while the French had a very serviceable machine in the Nieuport 11.

Development in aircraft and flying skills was accompanied by similar progress in tactics. Close formation flying proved a safer method of venturing into enemy territory than single, gallant forays. This was particularly important for photographic reconnaissance missions; there was little point in taking a number of extremely important photographs if the reconnaissance aircraft was shot down before the pilot could return with them.

Air battles made a refreshing contrast to the massed attacks below and were often watched and cheered by the infantry. The aviator bore a close resemblance to the knight who jousted in former times: his fate depended on a combination of his own skills and the endurance of his mount. His life was likely to be short, but perhaps not as short as if he had remained with the army on the ground instead of volunteering to fly. Eventually a macabre form of league table was produced to show who was the most successful pilot in terms of kills. It was headed by the German Manfred von Richthofen with eighty kills. He was followed by René Fonk of France, with seventy-five, Edward Mannock of Britain with seventy-three and William Bishop of Canada with seventy-two.

Air power added an entirely new dimension to warfare. Firstly, it showed civilians that even if they never left their own neighbourhood they could be killed, maimed or made homeless just as easily as people in the areas where the armies were actually fighting. This fact gave an added edge to the pacifists and appeasers, who now gained support from many who cared little for the moral objections to fighting but a lot for the possibility that they might themselves become involved. Anything which reduces the will of a nation to fight, whether it is the sight of the wounded or the skilful propaganda of the aggressor, is an addition to the latter's firepower. Civilian morale is an important factor in the will to resist, for combatants fight better if they feel they have the support of those at home, but it is particularly important if production in factories is to continue. In the early days of bombing, the workforce in many factories engaged in essential war production downed tools and took shelter if a warning of approaching enemy aircraft was given; later, the management ceased to issue warnings but merely turned up the music.

Secondly, air power knew no frontiers. Attacks on shipping, which had previously thought itself almost immune, came as a considerable surprise to crews, which had begun to think they were both invincible and invulnerable. Crippled boats which had run the gauntlet of submarine attacks and limped into harbours were dismayed to find that they were still at risk. Formerly secure hiding places up rivers were easily spotted from the air. Aircraft reconnaissance could now make nonsense of tactical swoops by shipping, and the 'secret' arrival of reinforcements in a battle zone could be reported immediately, thus robbing the planned attack of any element of surprise. By the end of the war the all-round capacity of aircraft had increased enormously: speed, ceiling, flight endurance and armament were almost double those of the early models. The British Sopwiths and the French Spads were now more than a match for the German Albatrosses. But the most alarming aspect of the air war was what it presaged for the future. 'Mad scientists' (in other words those with more imagination than usual) had predicted that air power might be able to knock out the heart of a belligerent country and force it to immediate surrender; their views were greeted with scorn. However, on 13 June 1917 – soon after Britain had been greatly cheered by the successful mining of the Messines Ridge in Belgium – twenty German Gotha G IV bombers (aircraft with a top speed of 80 mph) flew to London in daylight and dropped seventy-two bombs on the city: 162 people were killed and 432 wounded; the Germans lost six aircraft. The British reacted by forming a bomber force to give the Germans a taste of their own medicine, and from October 1917 onwards were steadily raiding German cities.

The year 1917 also saw the appearance of the first aircraft carrier, which was in fact a heavy cruiser adapted to provide a deck for taking off. At that stage receiving an aircraft back on board was not a feasible proposition, and the returning Sopwith Pup (equipped with floats) had to settle on the water and subsequently be hauled aboard. But, as the Second World War proved, 'from small acorns mighty oaks do grow'.

The fact that Germany was able to rebuild its air force in defiance of the Versailles Treaty was entirely thanks to Soviet help. Potential German pilots were trained at a special flying school at Lipetsk, while the grateful Germans allowed potential Soviet staff officers to attend their General Staff Training School in Berlin.

By the mid-1930s the future pattern of aerial warfare was beginning to take shape. The first of the American Boeings, the B.9, made its appearance and demonstrated that it could achieve the alarming speed of 188 mph. The Soviets, however, had meanwhile produced the Tupolev TB-3, a four-engined bomber which, though much slower than the B.9, could travel twice the distance; furthermore it could be used to deliver parachutists, whose possibilities in war had recently been demonstrated by the Italians on manoeuvres.

With these new and exciting toys at hand, it was obvious that the totalitarian countries would be itching to use them. Mussolini used his to bomb and gas Ethiopian troops who had no means of retaliation; then, with Germany, he took on a more difficult task in the Spanish Civil War. Soviet aircraft and pilots were happy to supply the necessary experience by supporting the other side, but in general the German and Italian air forces were successful enough to be misled into believing that they had discovered the correct tactical doctrines for a larger-scale conflict.

When the Second World War broke out in 1939 both Britain and Germany, unknown to each other, had radar. It was an invaluable addition to a nation's firepower, for it enabled an air force or navy to know from which direction an enemy aircraft was approaching. As a result interceptors no longer needed to waste time and fuel on lengthy, and often fruitless, patrols; instead they concentrated themselves on, and above, the path of the oncoming bombers.

Aircraft were now two to three times as fast as their predecessors of the First World War. Blenheim, Wellington, Breguet, Heinkel, Tupolev and Mitsubishi bombers could now drone towards their targets at an average speed of 250 mph. However, they would be unlucky if they were intercepted by Hurricane fighters flying at 320 mph or Messerschmitt 109Es at 357 mph. Nevertheless, in spite of the vulnerability of unescorted bombers, many reached their destinations and caused extensive damage. The Germans used their aircraft with an inventiveness which constantly surprised the Allies. They employed them to knock out Allied airfields in surprise attacks; they made use of them to tow gliders full of troops; and in 1940 they captured the 'impregnable' Belgian fortress of Eben Emael, which was virtually the key to the frontier, by landing gliders on top of it. They used bombers to disrupt lines of communication, and fighters to machine gun and harass civilians, thereby spreading panic. Airborne terror became a weapon in its own right; the screaming bomb, mentioned earlier, was part of it. All this use of aerial firepower enabled the Germans to destroy the French will to fight. It was, however, a two-edged weapon, for though the German victory in 1940 proved that air power was of vital importance, it failed to reveal that it is not decisive under all conditions. Air power had proved decisive where targets were easily visible, as in France, Belgium, the Netherlands and Norway, and where the opposition was minimal; it would not prove to be so decisive when the target area had missile defence, or ground cover, as in Vietnam. It had not proved decisive against the RAF with its Spitfires and Hurricanes.

Between 1939 and 1945 there were many violent swings in the balance of firepower. The Japanese had won dramatic victories in the Pacific by means of surprise, highly trained troops, a monopoly of tanks and complete air superiority in Malaya, and a highly efficient Intelligence service. But by 1945 most of their aircraft were outclassed, their codes were being read, their carriers were being sunk by US submarines, and their calculations of enemy reactions were found to have been wrong.

Germany had begun the war very successfully with new tactics, tanks, aircraft and weapons, and seemed to be all-conquering. A few years later, however, she was experiencing the unthinkable. Little wooden aircraft from Britain, aptly named Mosquitoes, were penetrating deep into the Fatherland, initially to take photographs, later to drop bombs. The Mosquito Mk XVI was so fast, at 415 mph, that it needed no defensive armament, yet could reach Berlin and drop a bomb weighing 4000 lb. Bombs had increased dramatically in size: at the beginning of the war 500-lb and 1000-lb bombs were considered to be at the limit of practical possibilities, but within a few years Lancaster bombers could carry 12,000-lb bombs and eventually monsters of 22,000 lb.

From the USA came the B17 (Flying Fortress) and later B29, as well as the B24 (Liberator). They were escorted by P51 Mustang fighters. The Soviets produced the Ilyushin IL-2 Stormovik, which carried bombs but was also valuable for close support. It used rockets in large numbers and influenced other Allied aircraft to adopt the same tactics.

Parachute troops, although very useful, were not as decisive as had been hoped for. The problem with an attack by parachute troops is mainly one of supply. Even if many of the necessary supplies are sent down into the area where the attack is taking place, some are likely to go astray or be damaged in transit. During this war there was a distinct upper limit to the amount of heavy support material that could be delivered by aircraft, and it was therefore necessary for a successful parachute landing to be supported rapidly by a link-up from ground forces. After some encouraging successes earlier in the war, 1st British Airborne Division was scheduled to capture Arnhem, on the Lower Rhine, in September 1944. Unfortunately the dropping zone was a little too far ahead for the ground troops of 30 Corps to reach them and support the parachutists. Furthermore, the Germans had very strong ground forces in the area at the time and 1st Airborne, completely outnumbered and outgunned, suffered a disastrous defeat. Earlier in the war, in May 1941, a successful German parachute operation on Crete had shown that the price of victory might be too high. Fifteen thousand airborne troops went in by transport and gliders, but the losses were so great that the Germans never again attempted a similar operation. Supply of ground troops from the air was considerably more reliable than the reverse process. In Burma in 1944, 300,000 men on the ground were entirely supported by twenty-three squadrons of aircraft, each squadron consisting of up to eighteen aircraft. The workhorses of air transport during the Second World War were the Junkers 52, an ugly but efficient aircraft, and the Douglas DC3, otherwise known as the C47.

The most significant development at this period was the testing of jet aircraft. Turbojet propulsion had been shown to be feasible as early as 1931 by Frank Whittle, who at that time was a test pilot at the Marine Experimental Establishment at Felixstowe in Suffolk. However, the engine he designed was not put to use until 15 May 1941, when it was fitted to a specially built Gloster airframe; this long delay was caused by lack of the necessary research funds. Meanwhile two German designers, Ohain and Hahn, working with no knowledge of Whittle's experiments in the same field, produced a jet aircraft which flew for the first time on 27 August 1939 – a week before the Second World War began. However, several years passed before jet aircraft were considered reliable enough for normal military duties: the first in that category were the Gloster Meteor and the Messerschmitt 262 in 1944. Germany was working in some desperation on an advanced jet in 1945; it was claimed that if this had been produced in time it would have swung back the balance of air power, which was then in the Allies' favour.

There was considerable argument in high places as to how aircraft should be used during the war. Each service wished to have aircraft under its own command, but this was only possible to a very limited extent and for special purposes. The Royal Naval Air Service, or Fleet Air Arm, obviously had to have a certain number of its own aircraft, but the RAF was not prepared to see its valuable machines and pilots ordered about by generals or admirals who were not perhaps aware of how much they were asking. This said, the RAF went to enormous trouble in trying to meet requirements, however difficult and dangerous they might be.

There was also much debate about how resources should be used. Some thought that a continual strategic bomber offensive would be the quickest and cheapest way of bringing Germany to her knees. In the event it was neither, though undoubtedly it helped. The loss of aircrew was horrific, the number killed equalled that of subalterns killed in the trench warfare of 1914–18 – some fifty thousand. Strategic bombing was probably more effective in the Pacific theatre than in Europe. The most devastating air assault of the entire war was not that on Dresden, nor even the atomic bomb attacks on Hiroshima and Nagasaki, but the fire raid on Tokyo in March 1945. Three hundred and thirty-four B29s dropped a mixture of napalm and incendiaries on to the city, which was both densely populated and highly inflammable. The exact casualty figures are not known, but are estimated at 180,000. Similar attacks followed on Kobe, Osaka, Nagoya and other cities. Finally in August 1945 the nuclear age began. The bomb which fell on Hiroshima on 6 August weighed a mere 6 tons but it destroyed 4 square miles of the city and killed 72,000; many more were injured. Hiroshima was not, as has been alleged, a civilian target, but held the headquarters of five armies.

Nagasaki, whose turn came three days later, was a naval base. Here the Japanese gave the official casualty figures as 25,700 dead and 23,300 wounded; the fact that the figures were much lower than those at Hiroshima is explained by the fact that the latter is situated on flatter ground and there was nothing to protect any part of the city from blast.

It has been suggested that Japan was already beaten when the atomic bombs were dropped and that they were therefore both inhumane and unnecessary. Japan was undoubtedly beaten, but she was far from surrendering, and it was only the traumatic shock of these massive explosions that enabled the Emperor, who had no wish to see his people suffer further, to make the peace which many powerful Japanese did not wish for. There were still five million Japanese soldiers under arms and every Japanese on the home islands, men, women and even children, had been trained to fight in the event of invasion. Had that invasion been necessary, it is probable that over a million Allied soldiers would have been killed, all the prisoners of war held by the Japanese would have been killed as a safety precaution, and the final death roll among the Japanese themselves would have been vastly in excess of that caused by the fire raids and the atomic bombs.

Urgent military needs usually give an impetus to scientific research which ultimately benefits post-war civilians. Blood transfusions, 'wonder drugs' such as penicillin, and evacuation by helicopter were all urgently developed in order to preserve the lives of soldiers, sailors and airmen, and thus indirectly support firepower. Helicopters have subsequently saved the lives of thousands, particularly in disasters at sea, and long-distance transport aircraft, originally designed to deliver instruments of death, have now introduced unusual and exotic holidays to millions.

Air power was used in a very different way in 1948. During the period of what was known as the Cold War, the USSR sealed off the roads from the West to Berlin, which lay in Soviet controlled territory, though the city itself was divided into four sectors, each of which was administered by one of the four Allied powers: the Soviet Union, France, Britain and the United States. The Soviet aim was to cause the other Allies to give up the foothold in Berlin which they had retained because, as the historical capital of Germany, it was too important to be handed over as part of the Soviet controlled zone of

Germany. The other three powers responded to the threat by mounting a massive airlift which lasted fifteen months and flew in 243,313 tons of supplies in 277,264 flights. Faced with this response, the Soviets reopened the roads which had ostensibly been closed for repairs but still showed no sign of them. This was an example of potential firepower being used to maintain the status quo, not to change it. It was the basis of the future of deterrence, MAD – the mutually assured destruction referred to earlier.

However, this superpower confrontation did not prevent a smaller-scale war breaking out. North Korea, a Soviet satellite, invaded South Korea in 1950. South Korea appealed for help to the United Nations and a massive response was mounted by the United States and Britain; other countries also played a considerable part. Unfortunately, when the war appeared to have reached a stalemate, forthright moves and statements by General Douglas MacArthur alarmed the Chinese; they thought they might be invaded by US troops and therefore decided to take the preventative measure of invading South Korea before it could happen. Eventually a truce was called and a demarcation line between the two parts of Korea was agreed: it was to be monitored by representatives from both sides. During the fighting some notable air battles had been waged, in which the North Korean (Soviet) Mig-15s had many successes. The Mig-15, it was officially disclosed in 1982, was powered by an exact copy of Rolls Royce Derwent and Nene engines. The engines, fifty-five of them, had been exported to the USSR in 1946 after a Soviet trade delegation had complained to Sir Stafford Cripps, then President of the Board of Trade, that Britain was discriminating against the USSR in her trade policy. The British government of the day is said to have believed the engines to be obsolescent, but they can hardly have thought that of the twenty-five Nenes, which had only been used experimentally in Lincoln bombers and had not yet been supplied to the RAF. The USSR built eight thousand Mig-15s in five years; they could fly, climb and dive faster than any Western aircraft. However, the Soviets today hardly need to buy high technology. Most of it is there for the taking from public library files under the US Freedom of Information Act, which enables anyone, US citizen or not, to demand access to secret material. The Soviets must have a serious problem in collating and classifying the vast quantities of secret material they are able to obtain so easily from the West.

Since the end of the Korean War in 1953 there have been other limited conflicts in which the strength and limitations of air power have been demonstrated. Aircraft have been used for routine surveillance, an activity which has been greatly helped by developments in camera technology. The troop-carrying helicopter, first used in the late 1940s, has been invaluable and its companion, the helicopter gunship, has shown that these machines need not invariably have a passive role. Helicopters, being comparatively slow, are vulnerable to ground fire, particularly from missiles. The USSR used helicopter gunships very successfully in Afghanistan until the Afghans obtained various forms of missiles from sympathizers. Since the missiles arrived, Soviet activity has been much curtailed.

Combat helicopters have now firmly established themselves on the battlefield, whether over land or sea. France is very much to the fore in developing multi-purpose machines. Aerospatiale has produced the Dauphin, the Ecureuil, the Gazelle and the Puma; a number of the last-named have been sold to Argentina and were used in the Falklands War. Gazelles were used by Britain in the same conflict. Italy has two excellent helicopters in the Agusta A109A and the A129 Mangusta. The workhorse of the US Army, and of many other armies too, is the Bell Huey, closely followed by the Bell OH-58 Kiowa, the AH-1 Huey Cobra, the Hughes Apache and Defender, and the Kaman SH-2 Seasprite, though the best-known are probably the Sikorsky S-70 and the AUH-76. Perhaps the most versatile of all helicopters is the Westland Lynx, a joint Anglo–French product which performed admirably in the Falklands. Westland also produce two other excellent helicopters in the Scout and the Wasp.

The USSR, as may be expected, also has a formidable range of helicopters. The oddly nicknamed Hormone (the Kamov Ka-25) is a general-purpose machine, usually ship-based for anti-submarine warfare. Its companion, the Kamov Helix, is larger. Their large gunship helicopter is the Mil M1-24 Hind-D, a machine all too familiar in Afghanistan; the Hind-D's body has been strengthened with steel and titanium. The Soviets also have a range of attack helicopters in the Mi-8 and Mi-17 range, which are nicknamed Hip by NATO. Varieties of this model are used for electronic warfare and troop transport.

Helicopters are famous for the work they perform in rescuing victims of shipwreck or stranded mountaineers, but their contribution to survival on the battlefield is less appreciated outside the services. CASEVAC, as casualty evacuation is known, enables wounded men to be removed from the battlefield in time for appropriate medical and surgical aid. As this saves lives and enables many wounded men to return to combat, helicopters form a considerable addition to a nation's firepower, quite apart from their other activities as gunships, missile carriers, electronic warfare aircraft, transports and vehicles for tactical reconnaissance.

Another notable post-war development has been the VTOL (vertical take-off and landing) aircraft, which does not require a runway. These are aircraft of the 1980s, developed from the British Aerospace Harrier STOVL (short take-off vertical landing) aircraft, and have given fighters a new lease of life. The Sea Harrier is ideal for aircraft carriers, and like its land-based companion is highly manoeuvrable and versatile.

In spite of the vast range of ballistic and other missiles, large aircraft continue in service, albeit dogged by controversy. The US B52 Stratofortress has a wing span of 185 feet and weighs 470,000 lb when loaded. Its average speed is in excess of 500 mph and it can carry a range of bombs as well as the Harpoon cruise missile.

The B52 is now considered virtually obsolete, and its replacement is the Rockwell B-1. The Rockwell has a shorter wing span of 136 feet, but a greater loaded weight of 477,000 lb; it is also considerably faster, being capable of 600 mph loaded. The B-1 has taken a long time to come into service and now, even before the last one ordered has been delivered, it is considered obsolescent. Its replacement will be the Northrop B-2, the renowned Stealth bomber. This is undetectable by radar, heat, light or sound, but is apparently not as fast as the B52 nor as agile as some other aircraft. However, the fact that it can slip in, undetected and unseen, carrying cruise missiles and equipment for jamming enemy defences, makes it a very formidable aircraft indeed. It is claimed that it will give the USAF complete air superiority over the Soviet Union. It has neither tail fin nor tailplanes and is merely a flying wing – a huge flying triangle. Managing it is not easy, as its performance in the air is controlled by computer; nevertheless it has won the approval of those who have piloted it.

The Soviet Union has two aircraft in the strategic bomber class, both manufactured by Tupolev. The first, nicknamed Backfire, is a twin-engined aircraft with a 113-foot wing span and a weight of some

270,000 lb. Its normal cruising speed, with a full load, is approximately 550 mph but, in common with other aircraft of this type, it can double this if empty and only engaged on reconnaissance. The Soviets deny that this is a strategic aircraft, even though it has a range of 7500 miles and carries the AS.4 (Kitchen) missile with a nuclear warhead.

Blackjack is a much larger aircraft, with four engines, a wing span of 177 feet, a loaded weight of 683,000 lb and a range of 4500 miles. It can stay in the air for fourteen hours, which is four hours longer than Backfire, and it carries cruise missiles. Blackjack is only just coming into service, but as more are produced they will replace the Bear and Bison aircraft which have now been in service for a considerable time. Although Blackjack is the largest strategic bomber in existence, it is not necessarily the best. Undoubtedly the Soviets are aware of the US Stealth bombers and are probably experimenting along the same lines.

The French have a notable bomber in the Dassault-Breguet Mirage IV, which can be adapted for strategic bombing, missile carrying or reconnaissance. This is a twin-engined aircraft which can carry a nuclear bomb at 770 mph but, unloaded, can conduct air reconnaissance at 1454 mph.

France has a formidable range of tactical aircraft in the D-B Mirage range, many of which are in service with other nations: the 111 and 5 have had marked success with the Israeli Air Force. These are very well equipped all-round aircraft, with 30-mm cannon and capacity for a variety of bombs, and are easily recognizable by their tail-less delta wing shape. Their stable companions are the Mirage F1 (smaller), the 2000 (larger and better armed), the Super Etendard and the Dornier Alpha Jet; the last is a joint production with the West German firm of Dornier. The Super Etendards used by the Argentine Air Force had notable success in the Falklands War, when their Exocet missiles accounted for HMS *Sheffield* and the *Atlantic Conveyor*.

Britain has the British Aerospace Buccaneer and the Harrier, both of which are tactical attack and reconnaissance aircraft; the Hawk, the supersonic Lightning and the Sea Harrier, among others. During the Falklands War the Argentines used the Pucará, a close-support attack and reconnaissance aircraft designed and manufactured in Argentina, but without good effect.

The United States has a wide variety of multi-role fighter and attack aircraft which include the Cessna A37B Dragonfly, the A-10A

Thunderbolt, the F16 Fighting Falcon, the F111 (a versatile all-weather attacker), the Grumman EA-6 Prowler, the F14 Tomcat, the Lockheed F104 Starfighter, the McDonnell Douglas A-4 Skyhawk, the F4 Phantom II, the F15 Eagle, the F18 Hornet and the AV-8B Harrier II, which is a joint UK/US product. All these are excellent aircraft. In this same fighter-and-attack capacity comes the Israeli Dagger, which has had notable success in the various Arab–Israeli wars.

The Soviet Union can field the Mig 19 Farmer, the Mig 21 Fishbed, the Mig 23 and 27 Flogger, the Mig 25 Foxbat (with very powerful turbojets), the Mig 29 Fulcrum and the Mig 31 Foxhound. In this category come the Japanese Mitsubishi F-1 and the Chinese Q-5 Fantan.

In the light tactical fighter class come the US Northrop F-5, the F-20 Tigershark and the Rockwell OV-10. Finally, a most successful international development is the Panavia Tornado, a joint enterprise by Britain, West Germany and Italy, used by NATO.

It is, of course, impossible in a book of this length to describe the different fighter, attack and multi-purpose aircraft in any detail – to do full justice to these versatile machines would require a book of their own. They are by no means identical, but include many variations and innovations in their efforts to attain superiority over their potential rivals. There are of course many others: Sweden has a range of Saabs; France has the Sepecat Jaguar; China the J-8 Finback; Italy the Marchetti S-211; Yugoslavia the Soko group; Russia the Fishpot, Fitter, Frogfoot, Flanker, Fencer, Firebar and Forger. Not to be overlooked is the US Vought A-7 Corsair II. But it must be emphasized that these by no means cover the whole range.

In April 1987 Britain announced that it was increasing its air firepower by purchasing advanced medium- and short-range air-to-air missiles, which will be fitted to the Hawk and Tornado fighters. Each Tornado will probably carry between four and twelve missiles.

Nimrod Mark 3, which was cancelled in 1987, was an AEW (airborne early warning) aircraft. Instead six Boeing E-3 Sentry aircraft will be used for AWAC (airborne warning and control) purposes. The Boeing E-3, like the Grumman Hawkeye and Raven, and the Soviet Ilyushin 11–18 and Tupolev TU-126, is an electronic warfare aircraft with OTH (over the horizon) capability. One of the greater dangers in air warfare is the attack which comes in below normal radar level, perhaps in the form of a cruise missile. AWACs are designed to observe such moves electronically, and if this ability can be extended to an OTH capability, so much the better. All these aircraft have an unusual, often bizarre, appearance owing to the need to carry complicated electronic apparatus which must not suffer interference by the host aircraft. An AWAC aircraft keeps watch over an area of 300 miles' radius and monitors the movement of all aircraft and shipping, friendly or not. The movements of submarines are also kept under surveillance by British Westland Sea-King helicopters, by American Sikorsky S61s, by French DB Atlantiques, by Dutch Fokker F27 Maritimes, by US Lockheed Orions and Vikings and by Soviet Ilyushin 11-38 Mays and Mil Mi-14 Hazes. Japan has an interesting machine in the Shin Meiwa SS-2, a flying boat which can fly slowly enough to keep a submerged nuclear submarine locked into its sonar beam.

One might ask why, in these days of unlimited missiles, aircraft have not long since been consigned to the scrap heap. The reason is that they are almost invulnerable, perhaps even more so than nuclear submarines. The United States Strategic Air Command never has less than a third of its aircraft in the air at any time. If a surprise attack were to be made on the United States the incoming missile might destroy every ICBM in its silo, but even before that happened the great aircraft endlessly patrolling the skies would have received their instructions. They would proceed to take their revenge, thus ensuring that there would be no winners in a Third World War.

The nuclear powers are well aware of the inherent dangers of this situation and have taken numerous precautions to make sure that war does not break out by accident. Hotlines exist between the major countries, so that heads of state can immediately resolve any doubts about each other's intentions or actions; additionally, all those responsible for operating nuclear weapons have elaborate fail-safe procedures to make sure that no action is taken without the certainty that there is no alternative. In fact, the great powers are more concerned about the actions which might result from the decision of irresponsible states or terrorist groups than with what one of their peers might do. Nevertheless, they are still trapped in an arms competition which may well continue for years, in spite of a more commonsense attitude being expressed in 1987.

12

Conclusions: Firepower in a Nuclear Age

Sadly, one result of man's exploration of space has been that he has come to regard it as an extension of Battlefield Earth. Since the beginning of 'civilization' theatres of war have become steadily larger; whereas the fate of a nation could once be decided on a single field, as England's was at Hastings in 1066, battlefields have extended to whole countries, even continents, and now encompass the globe. With the ability to launch vehicles into space men opened up a new, even more appalling, dimension. This, inevitably, developed its own terminology.

In space there are two types of offensive (or defensive) weaponry. The first is the simple satellite, which may be exclusively used for communication and thus does not carry any form of weapon. Unarmed though this satellite is, it can form an essential component of an armed force on the ground. It can provide communications and, by carrying sophisticated cameras, act as a 'spy in the sky'. Because of these functions satellites become militarily important and may need to be destroyed. The anti-satellite weapons used for this purpose are known as asats (or A-SATS).

The second type of weapon is purely offensive. It is known as an ABM, which stands for anti-ballistic missile. This type of weapon is capable of considerable variation: it may be another rocket targeted to home on to one already in flight, it may be a laser beam, or it may be a dense cloud of small projectiles which will cut to pieces the missile in orbit. The ABM may include weapons fired from a space shuttle which remains in orbit. However, the value of a space shuttle as a weapon-launching platform seems debatable. Steady, in orbit, it could make the perfect target. A less vulnerable type of base in space, from which weapons could be targeted to places on earth, is the moon. Undoubtedly there must be other types of space weapons under development and these, being entirely original, may, repeat *may*, be the reasons for reported sightings of 'flying saucers' or UFOs (unidentified flying objects).

War in space may well be inhibited by some of the same reasons as those which affect war on earth. Deductions made from the results of high-altitude nuclear tests show that explosions which take place above the atmosphere are likely to produce an effect so widespread that the initiating power's satellites would suffer as much damage as those of its enemy. A nuclear explosion produces gamma rays which in turn create electromagnetic radiation. This sends a surge of destructive power through all electrical circuits and conductors within its range; in space that range is considerable. This phenomenon is known as the electromagnetic pulse. Whether anything on the ground would survive the experience of a war in space or the upper atmosphere seems doubtful. Such explosions, if ever they occurred, would have an effect far into the future. Space is already littered with debris from former launchings. Many of these orbiting objects are very useful for satellite communication, for they can be used to 'bounce back' radio waves, but the thought of them being joined by the radioactive debris from a nuclear explosion in space is daunting.

After a series of experiments by both the USA and USSR both sides felt that the existence of a strategic balance, by which each power's fears of an attack by the other were calmed, would immediately be upset if one side or the other successfully deployed an ABM system. As a result, in 1972 the USA and USSR signed an ABM treaty. Since that date the Soviets appear not to have kept their side of the bargain for, as Margaret Thatcher mentioned in the broadcast which she was allowed to make over Russian television in 1987, an ABM system, designed to destroy ICBMs in flight, has now been installed around Moscow and certain other Russian cities. The fact that US Intelligence was already aware of this had had no small effect on President Reagan's decision to press forward with the Strategic Defence Initiative (SDI), better known to the public as Star Wars. Whereas in the past the superpowers had managed to live in peace with each other under MAD (mutually assured destruction), it now seemed as if the stakes had been raised a few notches higher. This was not entirely the result of the ABM/SDI situation, but owed something to the fact that many nuclear warheads were now MIRVs, a system which allowed the warheads to be released at different times to different targets: to intercept all of these would, of course, be extremely difficult. Such warheads also allowed for error: if a ballistic missile from Russia was aimed at a target in the USA, it might or might not reach it. Rockets develop faults in flight or go off course: however, when a missile site is the objective of a MIRV, if one warhead fails to reach it another will probably succeed. Therefore the MAD theory is undermined by the fact that the destruction might not always be mutual; MIRVs could give an advantage to one of the potential belligerents.

With this in mind, strategic planners asked their scientists and engineers to devise a system which would also come under the acronym MAD, but this time with the letters signifying mutually assured deterrence. Clearly, any system of complete deterrence would be enormously, perhaps prohibitively, expensive, even supposing it could be found. Equally clearly President Reagan believed that such a system was a practical proposition and wished to press ahead with it. It would, of course, require great flexibility and ingenuity, not to mention a massive underpinning by the electronics industry. Reagan said he was prepared to offer a similar system to the Soviets, but they regarded the invitation with some suspicion, wondering perhaps whether the system offered to them might have a few essential components missing. Soviet hostility to the Star Wars project probably had some link with an awareness that in the course of its development the USA might discover new electronic marvels which would be economically, as well as militarily, well out of Soviet reach. The fact that at the time of writing the superpowers might be on the brink of beginning to trust each other probably represents a more valuable breakthrough than any further agreements of the SALT (Strategic Arms Limitation Treaty) variety.

Unfortunately, ICBMs and their younger brothers are not usually as simple as this very basic description makes them sound. Most of them carry an interesting array of decoys to encourage the defender to attack the wrong target. However, these would avail them little if the ICBMs were confronted with the deterrent ABMs almost immediately after leaving their launching pads. In order for the deterrent ABMs to do this, they would need to be in orbit, not back with their owners. To make the defence even more effective, it would have to provide a series of barriers: a single barrier might not be enough, and seven is the figure now under consideration. Such defences would deploy a number of weapons in addition to the simple warhead which would collide with the approaching missile. The most important of the new range of weapons to be used in an ABM system are based on lasers.

'Laser' is an acronym, standing for light amplification by stimulated radiation. The word 'light' is possibly misleading, as indeed is 'amplification'. Lasers produce not visible light, but an infra-red invisible light. In simple terms, a laser is created when various liquid, gaseous or solid materials are stimulated between two aligned mirrors; the stimulation occurs electronically and affects the atoms. A laser beam has a multiplicity of uses: its energy can be concentrated to a point fine enough to be used in optical surgery, to cut diamonds, to bore holes in steel plate, to weld, to play recorded music and so on. The key to all these abilities is the fact that light is a form of radiant energy which creates electromagnetic waves. Laser beams travel at the speed of light – 186,000 miles per second – and can therefore pick up an oncoming missile before it has travelled very far. But the application of lasers as weapons, for instance to burn a hole in the armour of an enemy tank, is not easy. Producing a laser beam requires considerable electrical power, and this is seldom available on a normal battlefield. Chemical lasers seem to be a development of the future. If an infra-red beam is produced from hydrogen and fluorine it could burn a hole in an ICBM if it reached it. But even in 1987 lasers are clearly weapons in which both sides are investing much confidence and money. Reports have been received that Soviet spy ships are using laser beams to discourage close observation by interested aircraft and shipping.

Particle beams are another form of weaponry which is likely to be developed by Star Wars. These seem to be a near approach to the 'death ray' which has long been a favourite subject of science fiction writers. Particle beams differ from lasers in that, instead of using light, they focus a beam of sub-atomic particles. If the target was caught by a particle beam, gamma rays would be produced in it. The particles are usually electrons, though they could be protons: the electron is the smallest charged particle, while the proton and neutron have 1800 times its mass. The Soviets are known to have been experimenting – apparently with some success – but the fact that President Reagan has hinted at projects which will make nuclear bombs obsolete suggests that the Americans are also very active in this field.

The most controversial of the new range of weapons seems to be the neutron bomb, an enhanced radiation/reduced blast bomb for which the abbreviation ERRB is normally used. Where a normal nuclear bomb would kill by its explosive power, leaving a considerable degree of lethal radiation behind, the neutron bomb is designed to kill by neutron and gamma radiation used in a controlled manner. The effect of this new weapon would be to leave buildings, even tanks, intact but to kill people. Its range would not be great – perhaps 1000 yards – and the contamination would disappear within hours. The Soviets, anxious to discourage American progress on this weapon while working assiduously on their own version, spread the story that the neutron bomb was essentially a capitalist weapon: it did not matter how many people were killed provided property remained intact. Information about the present potential of the bomb is not disclosed, but it is thought not yet to be capable of being effective other than at short range on the battlefield. The thought of its being used to destroy cities conjures up those eerie fantasies, often found in children's fiction, of whole cities being deserted because all their inhabitants had been lured away and killed. US progress on the neutron bomb was delayed in 1978, when President Carter postponed it. Carter had a simple faith in the peaceful nature of the Soviet Union and looked forward to establishing a warm friendship between it and the USA; his illusions about that peaceful nature were, as he told the world, devastatingly shattered when he heard one morning that pacific intentions had motivated a Soviet Army to invade Afghanistan. When Ronald Reagan became President he announced that neutron bombs would be produced and stockpiled.

The unusual feature of the neutron bomb is that, although it is a hydrogen bomb, it has a very limited fall-out. A standard hydrogen bomb would produce a huge explosion which would scatter radioactive material over a wide area, in the process contaminating material objects, perhaps dust, which would take a long time to settle. By contrast the radiation in the neutron bomb is limited to a narrow area. Its drawback, from a military point of view, is that it often takes time to kill. Those contaminated could fight on, and with the knowledge that they were already doomed would make formidable opponents.

At the other extreme are bombs which specialize in producing a heavy degree of contamination. The cobalt bombs, which are nuclear bombs coated with cobalt, are designed to produce maximum contamination over a long period. These are essentially revenge weapons: they cannot be neutralized and would remain active for decades.

It is clear that nuclear war would wipe out most of the population of the earth – even those who lived far from the centre of the conflict – because air and water supplies would become so contaminated that survival would be improbable, if not actually impossible. However, from a study of the firepower available for conventional war it is also clear that a war between the superpowers involving non-nuclear weapons alone would make large parts of the globe uninhabitable. The destruction caused by non-nuclear weapons in the Second World War was not cleared up for at least a decade, and only then because America, whose industry had been untouched by the fighting, had enormous spare capacity to invest in the rebuilding of Europe and Japan. In a possible future combat the USA, and therefore the rest of the world, would not be so lucky; the increased range of aircraft and submarines, the greater power of explosives and the possibility of new destructive agents would make sure of that. In any event, it seems unlikely that even the industrial and financial might of an unravaged United States would be able to make a rapid restoration of a Europe in which a scorched earth policy had, wittingly or unwittingly, been used.

In fact the concept of a tidy little war, in which all the combatants agreed to avoid the use of nuclear, chemical or bacteriological weapons, is unrealistic. There are now so many nuclear power stations (mostly larger than Chernobyl) in the world that a proportion of them would certainly be damaged. Radioactivity would therefore poison wide regions, although this would not be the intention of the belligerents. Chemical or bacteriological warfare agents could be released in the same way. In 1943 an American transport, the SS *John Harvey*, was in Bari harbour, where it had just arrived with a quantity of stores for the US forces in Italy. Among them was a consignment of 100 tons of mustard gas shells, only to be released if the Germans, whose possession of large stocks of gas was well known, decided to use any against Allied troops. Unfortunately for the crew of the *Harvey*, a German aircraft arrived and bombed their ship, as well as several others. Some of the gas shells were damaged and leaked their contents into the harbour. There was very little smell, but within a few days eighty-three of the sailors were dead from mustard gas poisoning; 540 others suffered damage which was dangerous but did not prove lethal. With this precedent, it is not difficult to imagine what would happen if a large store of the latest chemical or bacteriological weapon was accidentally bombed. One thing is certain: the attacker would suffer as much as the attacked. Disease knows no frontiers.

By the same token, the USSR would suffer equally if ever she used nuclear weapons in Europe. The prevailing wind in the upper atmosphere is easterly. As the nuclear bombs exploded, radioactive clouds would rise high into the air and, taken by the prevailing wind, gently release their deadly burden over Soviet territory.

However, while we may discount the possibility of this type of warfare between superpowers – which would effectively destroy civilization, and perhaps all life on earth – there is no reason to suppose that lesser conflicts, such as the Arab–Israeli wars, the Falklands War, the Iraq–Iran War or even the Korean War, will not continue to be fought. This being so, two further aspects of firepower remain to be discussed: one is motivation, the other is espionage, and they remain as valid an element of firepower in the nuclear age as at any other time in history.

History offers many examples of battles being won by soldiers whose weapons were inferior to those of the army they defeated. Sometimes victory was achieved because the winners had commanders whose tactical skills were superior to those of the opposition: they chose their battlefields carefully and only began fighting when victory was certain. Sometimes the defeated side was already in a hopeless tactical position, pushed back on to marshy ground and with no space for manoeuvre or eventual escape. Sometimes they were beaten before the fight began because their opponents had a reputation for invincibility: in such circumstances only the coolest and most charismatic commander can snatch victory from defeat. Motivation may stem from the reputation of a commander such as Gustavus Adolphus of Sweden, Frederick the Great, Napoleon, Marlborough, Cromwell, Patton, MacArthur, Auchinleck or Slim.

Great battlefield commanders are frequently men who have suffered defeat but learned from their setbacks. The fact that they are fighting again, this time with a stronger, better-trained and better-armed force, often seems a pointer to victory. The test of a really good commander is to recover from defeat and to display flexibility in tactics. It is not possible to say whether Montgomery was a great commander, because he never gave battle unless he had overwhelming superiority; he was not, however, lacking in personal courage, and his training methods and ruthlessness with inefficient subordinates suggest that he was, to say the least, extremely capable.

Charisma (derived from the Greek *kharisma*, meaning favour or grace) is the word nowadays applied to the quality of being able to inspire others. Kitchener was a charismatic figure: his face on a poster motivated thousands to enlist, and there is a well-documented story of a young soldier being found dead on a First World War battlefield clutching a bloodstained photograph of Kitchener. Bernard Martin, a junior officer who fought in the murderous Battle of the Somme, related in his autobiography, *Poor Bloody Infantry*, how he had saluted General Douglas Haig, not nowadays considered a charismatic figure. Of Martin's platoon of thirty men, only himself and four others were still alive.

> We were an exhausted remnant, torn uniforms, rifles slung over shoulders anyhow, puttees slack, but all remarkably content to be alive – . . . When he was close I saw a man, stern-faced, expressing no emotion, riding slowly, for he left it to one of his Staff Officers to acknowledge the salute of our platoon as he passed. [But] I had saluted Sir Douglas Haig, Commander-in-Chief of the British Army, and was proud to have done so, as I think my four men were proud. Something to tell them at home one day.

He was of course saluting Haig the Commander-in-Chief rather than Haig the man. In the Army recruits are told: 'You are saluting the rank, not the man'. However, high rank itself is a charismatic factor.

The essential qualification of leadership in battle is the ability to convey, usually without words, the idea that the commander – whether a field marshal presiding over armies or a lance-corporal in charge of six men – is not ordering his men to do anything he would not more willingly do himself. There have inevitably been exceptions, but in general through the ages officers – however vain, foppish and selfish in other ways – have always accepted that responsibility in their position. The obligation of an officer or non-commissioned officer is to lead his men in action to victory. If his subordinates object to obeying his orders and do not wish to climb out of a trench and advance to almost certain death, he may shoot them, but victories are not won by shooting one's own side, and therefore his leadership is the all-important factor.

Courage must be taken for granted in a soldier, though his superiors must always be alert for any signs that it might be deficient or declining. It is no use supplying soldiers with the latest weapons if they abandon them and run away as soon as the bullets begin to fly. In 1987 many Libyan troops who had invaded the African state of Chad abandoned their sophisticated new Soviet-supplied equipment without firing a shot. They were not necessarily cowardly, for there is no reason to suppose that, given a different motivation, they would not have fought with the same scorn for death in battle as their contemporaries have displayed in other Middle Eastern wars. But on this occasion their motivation was inadequate.

Military history abounds with stories of fights to the last man, for patriotic, religious or purely traditional reasons. Sometimes these heroic last stands are unnecessary in their military context, but the fact that they occur means that the army to which the soldiers belong has a degree of motivation which is in itself a form of firepower.

Destroying another country's will to fight has always been an essential aspect of firepower. The similarity of methods for creating and destroying morale is remarkable. In the Middle Ages almost invariably each side had its cause blessed by priests and soothsayers who predicted victory. Sacred relics were often carried into the field; Edward I, known as the Hammer of the Scots, suggested that when he died his bones should be carried with all English armies marching into Scotland. His son, who disregarded this instruction, was heavily defeated at Bannockburn. The story was spread that angels had fought on the British side at the Battle of Mons in 1914 – but apparently they did not do so for long enough to ensure more than a temporary victory. More tangible assets were weapons which emitted a terrifying noise. Whether the walls of Jericho fell down literally at the sound of the trumpet seems open to doubt, but the arrival of a formidable force accompanied by terrifying blasts on trumpets and horns can be very disconcerting. The terror missile spans the centuries. From prehistoric times bowmen have appreciated that holes drilled in the shaft of an arrow can cause it to descend with a bloodcurdling howl; the Germans used the same technique with some of their bombs in the Second World War. The sound of bombs falling from aircraft is disconcerting enough: when some of them screech, like demons are alleged to do, the effect is considerable. In the Korean War of 1950–3 the Chinese launched mass attacks to the accompaniment of bugles and whistles.

Skilful propaganda or espionage also play their part in the psychology of war. In the 1930s the Nazi government gave financial and other encouragement to businessmen to enable them to purchase various French newspapers. Financial control led to editorial control, and articles reflecting the strength of the new Germany and the weakness of France, the folly of war and the importance of pacifism, appeared regularly in the French press. As a result, France was beaten in 1940 before the fighting had properly begun. Today the same process is active in the Western world, but the approach is more refined. Dedicated Marxists have sought to create the impression, usually in universities and schools, that Western civilization is essentially rotten and that salvation can only come from Communist governments which will be assisted by a benign Soviet Union. The benign quality of the Soviet Union's vast standing Army, Navy and Air Force and stocks of nuclear missiles is only mentioned when it is necessary to indicate that, if war ever came, the Soviets would win effortlessly.

In the 1930s, belief in all that the Soviet Union represented caused well-educated, intellectually capable young men to apply themselves to the acquisition of positions of responsibility and trust in the British government; once there, they proceeded to betray their country by passing its secrets to potential enemies. Blunt, Philby, Burgess, Maclean and no doubt others did their native country infinite harm. Klaus Fuchs, who had been accepted into Britain as a refugee from Nazi Germany, went on to betray the people who had received him, for he gave away their most vital secrets.

The situation does not require a moral judgement, but it undoubtedly requires a practical response. Traitors endanger the lives of innocent people. Soldiers die because the secrets of their weapons or tactics have been sold or given to the enemy. The penalty for traitors in peacetime in the West should be the same as it is in the East.

The more important spies may hold the key to whether we all have a future or not. Nobody with an iota of sense can believe that Britain, France, West Germany or the United States has the slightest wish to attack the Soviet Union. The USSR could disarm unilaterally and would still be completely safe. The same, alas, could not be said for the West over the last forty years, although there are more hopeful aspects for the future. The world of mutually assured destruction is the world of fearful apprehension. Each of the potential enemies is afraid that the other may gain a sudden over-riding advantage which might tempt a pre-emptive strike. We in the West know that this would never happen, but it is possible that a traitor could disclose to a potential enemy a vital discovery, perhaps one derived from Star Wars. Such information could give the impression that a pre-emptive strike was essential. This possibility, combined with the fact that traitors have caused infinite damage to the security of NATO countries in the last forty years, makes one wonder whether a less humane approach to traitors and their sponsors is now an obligation in the West.

A *Guardian* book review dated 10 July 1987 quoted the following statistics:

> Every day, two new nuclear missiles on average are now being deployed. Every two days, therefore, the world's arsenals gain a firepower equal to the whole of that used in the Second World War, and they already contain three thousand times that firepower.

Much of the present-day speculation about what would happen in a nuclear war is made invalid by the fact that no one can predict whether or not warring nations would resort to such weapons, and if so, how soon. For nations to use nuclear weapons against each other has been compared to two men fighting with grenades in a telephone box. One thing, however, is certain. If nuclear weapons are ever used in a future war the results will be incalculable. Whatever optimistic predictions may be made about the possibility of being able to advance over a nuclear battlefield within a short period after the use of the weapon, it does not alter the fact that the contamination cannot be anything but long-lasting. Britain would hesitate to use nuclear weapons because survival in a nuclear war would be unlikely for the inhabitants of a small island. However, this does not mean that

Britain should not possess nuclear weapons. The threat to Britain is not nuclear: it is the probability that in a conventional war the huge Soviet armies and air forces, not to mention a navy for which the USSR can have no conceivable defensive role, could combine to overwhelm Europe. But nothing is, of course, certain in war and it is highly likely that, once the Soviet war machine became as extended as it would need to be to overwhelm Europe, it might disintegrate for no more damaging reason than logistics.

If for no other reason, the stupendous cost of the arms burden may lead to mutual reduction. The suspicions of the USSR about the intentions of the West, many of which were based on events of the 1920s and the West's distrust of the alien Soviet system, seem to be diminishing. The superpowers may draw together to combat other problems, of which the terrorist threat by minority groups is not the least. The price of peace will, undoubtedly, continue to be eternal vigilance. We have now had freedom from a major war for forty years. There is no reason why the world as we know it should dissolve in a nuclear holocaust. The history of firepower has been the evolution of weapons which eventually attain parity. From the point of the ultimate deterrent we may be able to withdraw to a lower platform of security and build stronger bridges of understanding and trust between nations. It will not be easy. The world abounds with fanatics and self-seekers. But in the closing years of the twentieth century we may be at last moving towards a saner and safer world.

Further Reading

Archer, O., *Heavy Automatic Weapons*. Jane's 1978
Barnaby, F., *The Automated Battlefield*. OUP 1987
Batchelor, J., *Land Power*. Phoebus 1979
Batly, P., *The House of Krupp*. Secker & Warburg 1966
Bonds, R., *Modern Weapons*. Salamander 1985
Carman, W. Y., *A History of Firearms to 1914*. RKP 1955
Chant, C., *Warplanes*. Winchmore 1983
Friedman, N., *Submarine Design and Development*. Conway Maritime 1984
—— *Carrier Air Power*. Conway Maritime 1981
Geissler, E., *Biological and Toxin Weapons Today*. OUP 1986
Hackett, General Sir John, *The Third World War*. Sidgwick & Jackson 1982
Hardy, Robert, *The Longbow*. Patrick Stephens 1976
Hastings, Max, *Bomber Command*. Michael Joseph 1979
Isby, D., *Weapons and Tactics of the Soviet Army*. Jane's 1981
Johnson, N. L., *Soviet Military Strategy in Space*. Jane's 1987
Macksey, K., *The Tank Pioneers*. Jane's 1981
—— *Tank Warfare*. Hart-Davis 1971
Messenger, C., *Anti-Armour Warfare*. Ian Allen 1985
Morris, E. and others, *Weapons and Warfare of the 20th Century*. Cathay 1975
Oakeshott, R. E., *The Archaeology of Weapons*. Lutterworth 1983
O'Neill, R. *Guide to Weapons of the Modern US Army*. Salamander 1983
Peterson, H. P., *The Book of the Gun*. Hamlyn 1983
Reid, A., *A Concise Encyclopaedia of the Second World War*. Osprey 1974

Rose, S., *Chemical and Biological Warfare*. Harrap 1968
Rust and Brassey, *Weapon Technology*. London 1978
Shepperd, G. A., *Arms and Armour 1660–1918*. Hart-Davis 1971
Sweetman, John, *Operation Chastise*. Jane's 1982
Warner, Philip, *Alamein*. Kimber 1977
—— *The D Day Landings*. Kimber 1980
—— *Invasion Road*. Kimber 1980
—— *The Medieval Castle*. A. Barker 1971
Young, Brigadier Peter, *The British Army 1642–1970*. Kimber 1967

Index